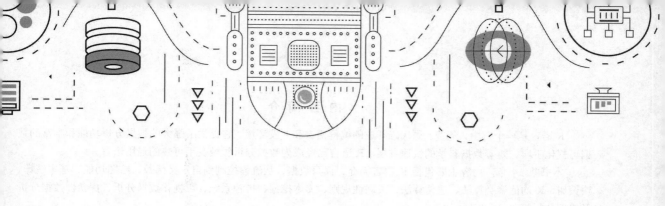

R语言
数据分析
从入门到实战

李仁钟 著

U0286241

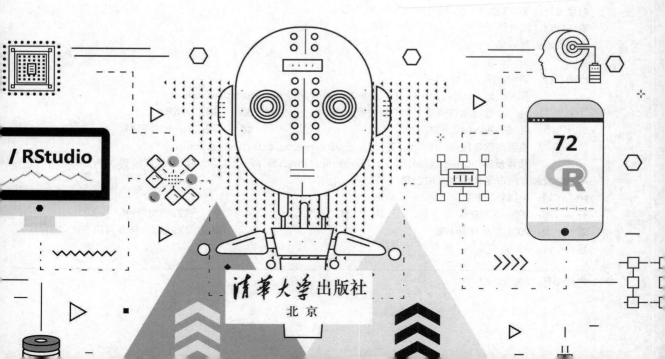

清华大学出版社
北京

内 容 简 介

R语言是一个自由、免费、源代码开放的编程语言和开发环境，它提供了强大的数据分析功能和丰富的数据可视化手段。随着数据科学的快速发展，R语言已经成为数据分析领域炙手可热的通用语言。

本书共14章，内容主要包括R语言简介、读写数据、从流程控制到自定义函数、绘图功能、基本统计、决策树、K均值聚类算法、遗传算法、关联性规则、文本挖掘、推荐系统、可视化数据分析、探索性数据分析及案例分析等。

本书内容通俗易懂，案例丰富，实用性强，特别适合R语言的入门读者和进阶读者阅读，也适合数据分析人员、数据挖掘人员等其他数据科学从业者阅读参考。

本书封面贴有清华大学出版社防伪标签，无标签者不得销售。
版权所有，侵权必究。举报：010-62782989，beiqinquan@tup.tsinghua.edu.cn。

图书在版编目（CIP）数据

R语言数据分析从入门到实战/李仁钟著. —北京：清华大学出版社，2021.7（2024.5重印）
ISBN 978-7-302-58634-0

Ⅰ. ①R… Ⅱ. ①李… Ⅲ. ①程序语言－程序设计 Ⅳ. ①TP312

中国版本图书馆CIP数据核字（2021）第142468号

责任编辑：夏毓彦
封面设计：王　翔
责任校对：闫秀华
责任印制：刘　菲

出版发行：清华大学出版社
网　　址：https://www.tup.com.cn，https://www.wqxuetang.com
地　　址：北京清华大学学研大厦A座　　邮　编：100084
社 总 机：010-83470000　　邮　购：010-62786544
投稿与读者服务：010-62776969，c-service@tup.tsinghua.edu.cn
质量反馈：010-62772015，zhiliang@tup.tsinghua.edu.cn

印 装 者：三河市龙大印装有限公司
经　　销：全国新华书店
开　　本：190mm×260mm　　印　张：13　　字　数：333千字
版　　次：2021年9月第1版　　印　次：2024年5月第5次印刷
定　　价：59.00元

产品编号：092702-01

前　　言

随着 R 语言（整合系统为 R 软件）的流行及普及，许多学者及专家转而使用 R 语言作为研究与开发的工具。R 软件有 Windows、UNIX、Linux 及 MacOS 等不同操作系统的免费版本，更有 1 万种以上的免费套件可供使用，所以学习 R 软件是很明智的选择。

本书前 5 章介绍 R 语言的基本操作及应用，第 6~9 章介绍各类学习算法，第 10~13 章详细介绍关联性规则、文本挖掘、推荐系统、可视化数据分析，第 14 章介绍探索性数据分析及案例分析。

作者是福州大学先进制造学院教授，发表论文 240 余篇，其中 SSCI、SCI、EI 等收录 80 余篇次，发表的论文多次获得最佳论文奖及优秀论文奖。本书是作者多年来从事教学的心血结晶，适合学习数据分析的读者阅读，本书范例中的程序代码也可以进行练习。本书以完全不懂 R 语言及数据分析的读者为对象，对于有意愿自我进修的读者而言，本书也是一本不错的入门参考书。本书的撰写虽力求完美，但难免会出现疏忽的地方，欢迎各位读者批评指正。

本书配套的源代码和教学 PPT 课件请用微信扫描下方的二维码获取，也可按扫描出来的页面提示把下载链接转到自己的邮箱中下载。如果学习本书过程中发现问题，请联系 booksaga@126.com，邮件主题为"R 语言数据分析从入门到实战"。

本书的出版感谢出版社编辑的鼎力协助，感谢福州大学先进制造学院领导们的支持，最后也感谢亲爱的家人的协助与支持。

李仁钟
2024 年 5 月

目　　录

第1章　R 简介 ... 1
1.1　R 软件介绍 ... 1
1.2　R 对象介绍 ... 4
1.2.1　向量 ... 4
1.2.2　数组 ... 5
1.2.3　矩阵 ... 8
1.2.4　数据框 ... 11
1.2.5　因子 ... 12
1.2.6　列表 ... 12
1.2.7　对象转换 ... 14
1.3　习题 ... 15

第2章　读写数据 ... 16
2.1　读取数据 ... 16
2.2　写入数据 ... 20
2.3　读写 RData 数据 ... 21
2.4　读取 SQL Server 数据库数据 ... 22
2.5　读写 Excel 数据 ... 23
2.6　习题 ... 23

第3章　从流程控制到函数 ... 24
3.1　条件执行 ... 24
3.2　循环控制 ... 26
3.3　函数 ... 29
3.4　习题 ... 30

第4章　绘图功能及基本统计 ... 31
4.1　高级绘图 ... 31
4.2　低级绘图 ... 34

4.3　交互式绘图 ·· 35
　　4.4　图形参数 ·· 37
　　4.5　基本统计 ·· 39
　　4.6　习题 ·· 44

第 5 章　数据分析和常用的包介绍 ·· 45
　　5.1　机器学习介绍 ·· 45
　　5.2　数据挖掘介绍 ·· 46
　　5.3　文本挖掘介绍 ·· 46
　　5.4　常用的包介绍 ·· 46

第 6 章　监督式学习 ·· 54
　　6.1　决策树 ··· 54
　　6.2　支持向量机 ··· 66
　　6.3　人工神经网络 ·· 70
　　6.4　集成学习方法 ·· 75
　　　　6.4.1　随机森林 ··· 76
　　　　6.4.2　提升法 ·· 76
　　6.5　习题 ·· 77

第 7 章　非监督式学习 ··· 78
　　7.1　层次聚类法 ··· 78
　　7.2　K 均值聚类算法 ··· 81
　　7.3　模糊 C 均值聚类算法 ·· 83
　　7.4　聚类指标 ·· 90
　　7.5　习题 ·· 92

第 8 章　演化式学习 ·· 93
　　8.1　遗传算法 ·· 93
　　8.2　人工蜂群算法 ·· 99

第 9 章　混合式学习 ·· 102
　　9.1　人工蜂群算法混合决策树 ··· 102
　　9.2　遗传算法混合人工神经网络 ·· 105

第 10 章　关联性规则 ··· 119
10.1　产生关联性规则并排序 ··· 121
10.2　删除冗余规则 ··· 126
10.3　习题 ··· 132

第 11 章　文本挖掘 ··· 133
11.1　使用混合分词并创建词频表 ··· 133
11.2　使用 tag 分词并创建词云 ··· 134
11.3　习题 ··· 136

第 12 章　推荐系统 ··· 137
12.1　Jester5k 数据集 ··· 137
12.2　MovieLense 数据集 ··· 140

第 13 章　可视化数据分析 ··· 142
13.1　导入数据 ··· 143
13.1.1　处理数据集 ··· 146
13.1.2　设置变量 ··· 147
13.2　探索及测试数据 ··· 147
13.3　转换数据 ··· 151
13.4　建立、评估及导出模型 ··· 153
13.5　习题 ··· 155

第 14 章　探索性数据分析 ··· 156
14.1　dplyr 数据处理库（包）··· 156
14.2　案例分析 ··· 165

附录 A　安装 R ··· 189
附录 B　安装 RStudio Desktop 和 rattle ··· 192
附录 C　R 语言指令及用法 ··· 197

第 1 章

R 简 介

　　R 是属于 GNU 系统的一款自由、免费、源代码开放的软件，它是一款整合型的数据分析软件、统计软件、绘图软件，同时也是一门程序设计语言。R 当初是由 Ross Ihaka 与 Robert Gentleman（1966）开发出来，与 AT & T 贝尔实验室 Rick Becker、John Chambers 与 Allan Wilks 等人所开发的 S 语言相似。R 有 Windows、UNIX、Linux 及 MacOS 等不同操作系统的版本。R 目前开发的核心团队是由世界各地不同的机构组成的，其网站地址为 http://www.r-project.org，在此网站上可查阅许多有关 R 的文件、相关书籍及信息。R 的应用领域包含数据分析、统计分析、数据挖掘、机器学习、推荐系统、文本挖掘及深度学习等。

1.1 R 软件介绍

　　读者可从 http://www.r-project.org 网站下载适合自己操作系统的新版 R 软件。安装 R 软件后，可在 http://www.rstudio.com/下载 RStudio。RStudio 是一个为 R 设计的集成开发环境。R 和 RStudio 在 Windows 操作系统上的安装步骤如附录 A 及附录 B 所示。

　　R 网站中提供了功能非常强大的工具集，我们可以从网站上安装相关应用的包（Package，或称为程序包，软件包），R 提供了 1 万个以上的免费包。当我们的计算机连接到因特网后，若使用 Windows 版本，则很容易通过"程序包"菜单选项来安装这些免费的程序包。可从该菜单中选择"加载程序包"选项来选择可用的包。在我们选择好想要的包之后，R 软件将下载所选用的包并自动进行安装。本书中的范例及操作皆在 Windows 操作系统下进行，如果我们要在 UNIX、Linux 或者 MacOS 上运行 R 软件，则可能需要进行相应的调整。我们也可以自行安装包，例如安装 C5.0 决策树包 C50（注意英文字母大小写是否相同），只需要在 R 提示符号">"后输入以下指令（注意：当提示符号为"+"时，表示程序正在执行中，或在等待未执行完成的指令）：

```
> install.packages("C50")
```

可使用以下指令来调用 C50 包中提供的函数：

```
> library(C50)
```

若要删除已安装的包，例如 C50 包，可使用下面的指令：

```
> remove.packages("C50")
```

R 软件是一种语法非常简单的表达式语言（Expression Language）。R 语言支持对象（Object），对象名称（变量）第一个字母须为英文字母或句点"."，若以句点为对象的第一个字母，则其后接的第一个字符不能为数字，例如.2iswrong 不能当对象名称。对象不需要事先声明，但对象名称中的字母大小写代表不同的对象，因此 X 和 x 是不同的对象名称。R 语言保留了一些标识符作为指令名称，即保留字，如 c 与 NA 等。R 语言可使用赋值（Assignment）表达式"<-"来进行赋值操作（也可以使用"="），例如：

```
# 赋值表达式
> x <- 10
> x
[1] 10

> X <- x^2
> X
[1] 100
> z <- sqrt(X)
> z
[1] 10
```

也可以通过对象名称的数据种类（属性，Attribute）来描述其对象的特性，也就是说，一个对象名称的作用取决于该对象名称的属性。所有的对象名称都有两个内在属性：数据类型（Mode）和长度（Length）。对象名称中的元素（Element）共有 4 种基本数据类型：数值（Numeric）、字符串（Character）、复数（Complex）和逻辑（Logical）。虽然也存在其他的数据类型，但是并不能用来表示数据，例如函数（Function）或表达式（Expression）。长度（Length）是指对象中元素的数量。对象的数据类型和长度可以分别通过函数 mode() 和 length() 得到。

```
# 对象的数据类型和长度
> x <- 10
> x
[1] 10

> mode(x)
[1] "numeric"

> length(x)
[1] 1
```

如果要在同一行中运行多个表达式，则可以使用分号";"隔开这些表达式，例如：

```
> x <- 10; y <- x^2; z <- sqrt(y)
> z
[1] 10
```

注释可以放在程序中的任何地方，从"#"号开始的行就是注释，例如：

```
#整数
> x <- 10
> x
[1] 10
> mode(x)
[1] "numeric"
> length(x)
[1] 1
#浮点数(实数)
> y <- 10.9
> y
[1] 10.9
> mode(y)
[1] "numeric"
> length(y)
[1] 1

#逻辑
> z <- T
> z
[1] TRUE
> mode(z)
[1] "logical"
> length(z)
[1] 1

#字符串
> a <- "Hello"
> a
[1] "Hello"
> mode(a)
[1] "character"
> length(a)
[1] 1

#复数
> z <- 4+2i
```

```
> z
[1] 4+2i
> mode(z)
[1] "complex"
> length(z)
[1] 1
```

1.2　R对象介绍

R是面向对象的程序设计语言，其常用对象有向量（Vector）、数组（Array）、矩阵（Matrix）、因子（Factor）、数据框（Data Frame）及列表（List）等。

1.2.1　向量

向量是由包含相同数据类型的元素所组成的，R程序中最简单的结构就是由一串有序数值构成的数值向量。假如用户要建立一个含有6个数值的向量V，且其值分别为10、5、3.1、6.4、9.2和21.7，在R程序中调用c()函数来实现。

```
# 向量
> V <- c(10, 5, 3.1, 6.4, 9.2, 21.7)
> V
[1] 10.0  5.0  3.1  6.4  9.2 21.7
> length(V)
[1] 6
> mode(V)
[1] "numeric"
```

也可以调用assign()函数来达到上述程序相同的结果：

```
> assign("V", c(10, 5, 3.1, 6.4, 9.2, 21.7))
> V
[1] 10.0  5.0  3.1  6.4  9.2 21.7
> length(V)
[1] 6
> mode(V)
[1] "numeric"
```

在某些情况下，向量的元素可能会遗失。当向量中的元素为缺失值（Missing Value）时，可给该元素赋予一个特定的值NA（需为大写）。

```
# 缺失值
> V <- c(10, 5, NA, 6.4, 9.2, 21.7)
> V
[1] 10.0  5.0   NA  6.4  9.2 21.7
```

我们可以使用中括号"[]"来访问向量中的特定元素。值得注意的是，在 R 语言中，向量对象中第一个元素的索引（Index）值默认是从 1 开始的，而不是从 0 开始。

```
# 访问向量中的特定元素
> V[2]
[1] 5
```

R 还提供 Inf、–Inf 及 NaN（Not a Number），而 NULL 是指对象的长度是 0。

```
# Inf -Inf NaN NULL 说明
>  V <- c(1,-2,0)
>  V/0
[1]  Inf -Inf  NaN

> V <- NULL
> length(V)
[1] 0
```

也可以使用":"创建向量。

```
# 创建连续向量并进行访问
> V2=1:10
> V2
 [1]  1  2  3  4  5  6  7  8  9 10

> V2[1]
[1] 1

> V2[2:4]
[1] 2 3 4
```

1.2.2 数组

数组可以看作是多维度的向量。例如，一个 3 维的数组 X 可以用 X[i,j,k]来指向其中的特定元素。假设数组 X 的维度向量是 c(3,4,2)，则 X 中有 3×4×2 = 24 个元素，这些元素依次为 X[1,1,1],X[2,1,1],...,X[2,4,2], X[3,4,2]。

假设 X 是一个包含 24 个元素的向量：

```
> X <- 1:24
```

可以调用 dim()函数来指定数组的维度（Dimension），让 X 变成一个 3×4×2 的 3 维数组，而 R 程序会按照列（Column）的方式排列：

```
# 调用dim()函数指定数组的维度
> dim(X) <- c(3,4,2)
> X
, , 1
```

```
            [,1] [,2] [,3] [,4]
    [1,]     1    4    7   10
    [2,]     2    5    8   11
    [3,]     3    6    9   12

, , 2

            [,1] [,2] [,3] [,4]
    [1,]    13   16   19   22
    [2,]    14   17   20   23
    [3,]    15   18   21   24
```

我们可以让 X 变成一个 4×6 的 2 维数组：

```
# 改变维度
> dim(X) <- c(4,6)
> X
         [,1] [,2] [,3] [,4] [,5] [,6]
    [1,]   1    5    9   13   17   21
    [2,]   2    6   10   14   18   22
    [3,]   3    7   11   15   19   23
    [4,]   4    8   12   16   20   24
```

要创建一个数组，也可以直接调用 array()函数，此函数的第一个参数用于指定数据向量，第二个参数用于指定数组的维度。假设要创建一个 3×4×2 的 3 维数组，指令如下：

```
# 调用array()函数来创建数组
> X <- array(1:24, dim = c(3,4,2))
> X
, , 1

            [,1] [,2] [,3] [,4]
    [1,]     1    4    7   10
    [2,]     2    5    8   11
    [3,]     3    6    9   12

, , 2

            [,1] [,2] [,3] [,4]
    [1,]    13   16   19   22
    [2,]    14   17   20   23
    [3,]    15   18   21   24
```

假设要创建一个 4×6 的 2 维数组：

```
> X <- array(1:24, dim = c(4,6))
> X
```

```
     [,1] [,2] [,3] [,4] [,5] [,6]
[1,]    1    5    9   13   17   21
[2,]    2    6   10   14   18   22
[3,]    3    7   11   15   19   23
[4,]    4    8   12   16   20   24
```

值得注意的是,以下指令会创建一个所有元素的值都是 0 的 3 维数组:

```
> X <- array(0, dim = c(3,4,2))
> X
, , 1

     [,1] [,2] [,3] [,4]
[1,]    0    0    0    0
[2,]    0    0    0    0
[3,]    0    0    0    0

, , 2

     [,1] [,2] [,3] [,4]
[1,]    0    0    0    0
[2,]    0    0    0    0
[3,]    0    0    0    0
```

也可以调用 rbind()和 cbind()函数来创建数组。rbind()表示把向量按行(Row)合并成一个数组,而 cbind()表示把向量按列(Column)合并成一个数组。

```
# 调用rbind()和cbind()函数来创建数组
> X1 <- c(1,2,3,4)
> X2 <- c(5,6,7,8)
> X  <- rbind(X1,X2)
> X
   [,1] [,2] [,3] [,4]
X1    1    2    3    4
X2    5    6    7    8

> X <- cbind(X1,X2)
> X
     X1 X2
[1,]  1  5
[2,]  2  6
[3,]  3  7
[4,]  4  8
```

1.2.3 矩阵

矩阵（Matrix）就是 2 维数组，要创建一个矩阵，可以调用函数 matrix()。在 R 4.0 以上的版本中，处理矩阵和数组更加一致，虽然在概念上矩阵只是 2 维数组，但是在 R 过去的版本中，在部分情况下，处理矩阵和数组对象的方法并不一样。在 R 的新版本中，矩阵对象将继承自数组类，从而消除了处理不一致的问题。

```
matrix(data = NA, nrow = 1, ncol = 1, byrow = FALSE, dimnames = NULL)
```

其中：

- nrow：表示矩阵的行数。
- ncol：表示矩阵的列数。
- byrow：表示矩阵中的数据是要按照行还是按照列（byrow=FALSE）的顺序排列。
- dimnames：可以给行或列命名。

```
# 调用matrix()函数创建2维数组
> X <- matrix(1:24, nrow=4, ncol=6, byrow=TRUE)
> X
     [,1] [,2] [,3] [,4] [,5] [,6]
[1,]    1    2    3    4    5    6
[2,]    7    8    9   10   11   12
[3,]   13   14   15   16   17   18
[4,]   19   20   21   22   23   24

> X <- matrix(1:24, nrow=4, ncol=6, byrow=FALSE)
> X
     [,1] [,2] [,3] [,4] [,5] [,6]
[1,]    1    5    9   13   17   21
[2,]    2    6   10   14   18   22
[3,]    3    7   11   15   19   23
[4,]    4    8   12   16   20   24
```

在 R 4.0 以上版本中，矩阵对象会继承自数组类，可调用 class() 函数来确认。

```
> class(X)
[1] "matrix" "array"
```

矩阵也可以调用 rbind() 和 cbind() 函数来创建矩阵。t() 是矩阵的转置（Transposition）函数，nrow() 和 ncol() 函数会分别返回矩阵的行数和列数。

```
# 调用 rbind()和cbind()函数来创建矩阵、转置矩阵
> X1 <- c(1,2,3)
> X2 <- c(4,5,6)
> X3 <- c(7,8,9)
> X <- cbind(X1,X2,X3)
```

```
> X
     X1 X2 X3
[1,]  1  4  7
[2,]  2  5  8
[3,]  3  6  9
> Y=t(X)
> Y
   [,1] [,2] [,3]
X1   1    2    3
X2   4    5    6
X3   7    8    9

# 调用nrow()和ncol()函数返回矩阵的行数和列数
> m <- nrow(Y)
> m
[1] 3
> n <- ncol(Y)
> n
[1] 3
```

若要显示矩阵 X 的第一列元素，则可执行如下指令：

```
# 显示矩阵整列的元素
> X[,1]
[1] 1 2 3
```

若要显示矩阵 X 的第二行元素，则可执行如下指令：

```
# 显示矩阵整行的元素
> X[2,]
X1 X2 X3
 2  5  8
```

若要显示矩阵 X 的第一行及第三行元素，则可执行如下指令：

```
# 显示矩阵部分行的元素
> X[c(1,3),]
     X1 X2 X3
[1,]  1  4  7
[2,]  3  6  9
```

若要删除矩阵 X 的第一列元素，则可执行如下的指令：

```
# 删除矩阵的一列元素
> X[,-1]
     X2 X3
[1,]  4  7
```

```
[2,]  5  8
[3,]  6  9
```

若要删除矩阵 X 的第二行元素,则可执行如下指令:

```
# 删除矩阵的一行元素
> X[-2,]
     X1 X2 X3
[1,]  1  4  7
[2,]  3  6  9
```

eigen()函数用来计算矩阵的特征值(Eigen Value)和特征向量(Eigen Vector)。

```
# 矩阵运算
> eigen(Y)
$values
[1]  1.611684e+01 -1.116844e+00 -1.303678e-15


$vectors
           [,1]        [,2]       [,3]
[1,] -0.2319707 -0.78583024  0.4082483
[2,] -0.5253221 -0.08675134 -0.8164966
[3,] -0.8186735  0.61232756  0.4082483
```

可用 "%*%" 运算符表示矩阵相乘:

```
> z <- Y%*%X
> z
   X1  X2  X3
X1 14  32  50
X2 32  77 122
X3 50 122 194
```

若要修改矩阵 z 的列名称,则可执行如下指令:

```
# 修改矩阵的列名称
> colnames(z) <- c("c1","c2","c3")
> z
   c1  c2  c3
X1 14  32  50
X2 32  77 122
X3 50 122 194
```

若要修改矩阵 z 的行名称,则可执行如下指令:

```
#修改矩阵的行名称
> rownames(z) <- c("r1","r2","r3")
> z
```

```
     c1  c2  c3
r1  14  32   50
r2  32  77  122
r3  50 122  194
```

1.2.4 数据框

数据框与矩阵的结构类似，因为两者的结构都是 2 维。然而，与矩阵不同的是，数据框可以在不同列中存在不同的数据类型，但同列的数据类型必须相同。数据框的每一行可视为一组观察值（Observation）或案例（Case），其变量名称是由每一列的名称来定义的。

可使用以下方式创建数据框：

```
# 创建数据框
> id <- c(1, 2, 3, 4)
> age <- c(25, 30, 35, 40)
> sex <- c("Male", "Male", "Female", "Female")
> pay <-c (30000, 40000, 45000, 50000)
> X.dataframe <- data.frame(id, age, sex, pay)
> X.dataframe
  id age    sex   pay
1  1  25   Male 30000
2  2  30   Male 40000
3  3  35 Female 45000
4  4  40 Female 50000
```

可使用以下方式取得或引用数据框中第 3 行第 2 列的元素：

```
> X.dataframe[3,2]
[1] 35
```

可使用列的名称取得或引用数据框中对应列的所有元素：

```
# 使用列的名称取得列元素
> X.dataframe$age
[1] 25 30 35 40
```

可使用以下方式取得或引用数据框中对应列的名称及元素：

```
# 使用列的名称取得列的名称及元素
> X.dataframe[2]
  age
1  25
2  30
3  35
4  40
```

R 程序提供了与 Excel 界面类似的编辑器来创建或修改数据框的值（见图 1-1）：

```
# 编辑或修改数据框的值
> edit(X.dataframe)
```

若确定要变更"修改后"的数据框，则需使用赋值运算符：

```
> X.dataframe <- edit(X.dataframe)
```

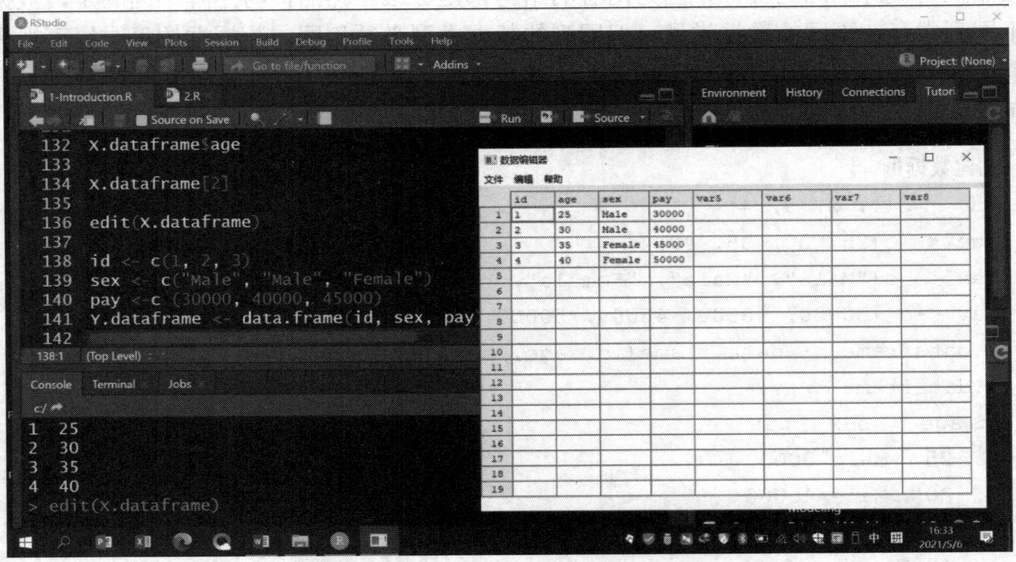

图 1-1　edit()函数

注意：必须先在要修改或编辑的字段上双击鼠标左键才可以开始修改。

1.2.5　因子

因子（Factor）是一种特别的向量，用于将同样长度的离散数据向量分组（Grouping）。字符串向量中每一个元素取一个离散值，因子有一个特殊属性称为水平（Levels），表示因子变量可以取得的所有离散值。一旦设置为因子，R 打印时就不会加上双引号。

可以调用 factor()函数来建立因子，其中 Levels 代表因子可以取得的所有离散值：

```
> sex <- factor(c("男", "女", "男", "男", "女"))
> sex
[1] 男 女 男 男 女
Levels: 女 男
```

1.2.6　列表

R 语言的列表是以有序序列（Order Sequence）构成的对象。列表的组成元素（Component，也简称元素）可以是异质（Heterogeneous）的对象，也就是说，各个组成元素的数据类型可以不相同。一个列表中的组成元素可以包括数值、逻辑、字符串、复数、向量、矩阵、因子及数据框等。

可以调用 list()函数来创建列表：

```
# 创建列表
> id <- c(1, 2, 3)
> sex <- c("Male", "Male", "Female")
> pay <-c (30000, 40000, 45000)
> Y.dataframe <- data.frame(id, sex, pay)

> gender <- factor(c("男","男","女"))
> Paul.Family <- list(name="Paul", wife="Iris", no.kids=3,
kids.age=c(25,28,30), gender, Y.dataframe)
> Paul.Family
$name
[1] "Paul"

$wife
[1] "Iris"

$no.kids
[1] 3

$kids.age
[1] 25 28 30

[[5]]
[1] 男 男 女
Levels: 女 男

[[6]]
  id    sex   pay
1  1   Male 30000
2  2   Male 40000
3  3 Female 45000
```

可以使用$来取得或引用列表中的某一个元素，例如欲获取 Paul.Family 中第 4 个元素，则可执行如下的指令：

```
# 使用$符号来取得或引用列表中的元素
> Paul.Family$kids.age
[1]  25 28 30
```

也可以使用双重中括号"[[]]"及索引值来获取或引用列表中某一个元素，例如要取得 Paul.Family 中第 4 个元素，可执行如下的指令：

```
# 使用"[[]]"符号来取得或引用列表中的元素
> Paul.Family[[4]]
[1]  25 28 30
```

若使用一个中括号"[]"及索引值可取得或引用列表中某一位置的组成元素及名称,则可执行如下指令:

```
# 使用"[]"符号来取得或引用列表中的元素及其名称
> Paul.Family[4]
$kids.age
[1] 25 28 30
```

注意,需使用双重中括号。若欲取得第二个孩子的年龄,则可执行如下的指令:

```
> Paul.Family$kids.age[2]
[1] 28
```

或

```
> Paul.Family[[4]][2]
[1] 28
```

1.2.7 对象转换

R 语言提供了多个函数用于不同对象间的转换,这些函数有 as.vector()、as.array()、as.matrix()、as.factor()、as.data.frame()及 as.list()。

创建数据框对象:

```
# 转换向量->数据框->矩阵->向量
> id <- c(1, 2, 3, 4)
> x <- data.frame(id)
> x
  id
1  1
2  2
3  3
4  4
```

把数据框对象转换为矩阵对象:

```
> matrix.x=as.matrix(x)
> matrix.x
     id
[1,]  1
[2,]  2
[3,]  3
[4,]  4
```

把矩阵对象转换为向量对象。

```
> vector.x=as.vector(matrix.x)
> vector.x
[1] 1 2 3 4
```

1.3 习题

(1) 使用 Repository 安装 C50 库（包），参考图 1-2。

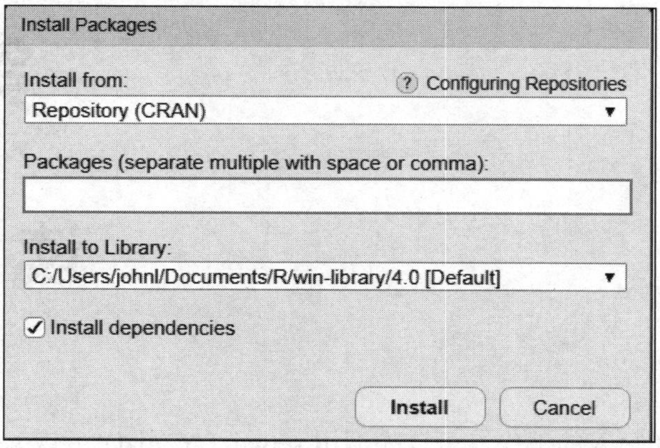

图 1-2　自动安装

(2) 使用 Package Archive File 安装 RGtk2_2.20.36.zip 库（包），参考图 1-3。

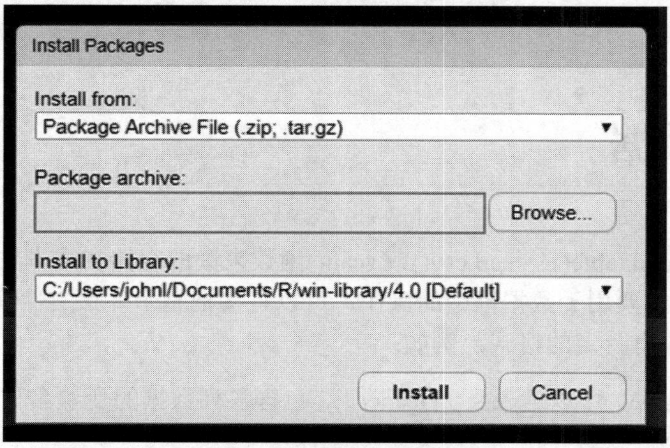

图 1-3　手动安装

第 2 章 读写数据

对于数据的读取和写入的工作，可以先调用 setwd（"C:/data"）或 setwd（"/home/R"）函数来切换工作目录，再调用 getwd()函数来确认当前工作目录。

```
> setwd("c:/")
> getwd()
[1] "c:/"
```

2.1 读取数据

R 语言常调用 read.table()、read.csv()或 scan()函数来读取存储在文本文件（ASCII）中的数据，read.table()函数主要用于数据框中的操作，可以直接把整个外部文件读入数据框对象。

外部文件常要求有特定的格式，例如：

（1）第一排（Line）可以是表头（Header），包含各列数据的变量名称（Variable Name）。不过表头也可以省略。

（2）其余各排是各行的值。

要读取在 R 的工作目录中的 X.csv 文件（见表 2-1），可执行如下的指令：

表2-1 X.csv文件

id	age	sex	pay
1	25	Male	30000
2	30	Male	40000
3	35	Female	45000
4	49	Female	50000

```
# 调用read.table()函数
> setwd("c:/")  # 设置工作目录

# encoding表示编码格式
> X <- read.table("X.csv",sep=",",header=TRUE, encoding="UTF-8")
> X
  id age    sex   pay
1  1  25   Male 30000
2  2  30   Male 40000
3  3  35 Female 45000
4  4  49 Female 50000

> X$age
[1] 25 30 35 49

> X[1,2]
[1] 25
```

注意，CSV 文件中的数据是用逗号分隔开的，所以加入 sep="," 来指定分隔符是逗号。若 header=FALSE，则使用默认的 V1,V2,…,V#来作为表头（header）的名称。

```
# 使用header参数要小心
> setwd("c:/")
> X <- read.table("X.csv",sep=",", header=F, encoding="Big5")
> X
  V1  V2     V3    V4
1 id age    sex   pay
2  1  25   Male 30000
3  2  30   Male 40000
4  3  35 Female 45000
5  4  49 Female 50000
```

也可以调用 read.csv()函数：

```
# 调用read.csv()函数
> setwd("c:/")
> X <- read.csv("X.csv", header=TRUE, encoding="GBK")
> X
  id age    sex   pay
1  1  25   Male 30000
2  2  30   Male 40000
3  3  35 Female 45000
4  4  49 Female 50000
```

```
> X <- read.csv("X.csv", header=FALSE, encoding="UTF-8")
> X
  V1  V2     V3    V4
1 id  age    sex   pay
2 1   25     Male  30000
3 2   30     Male  40000
4 3   35     Female 45000
5 4   49     Female 50000
```

可以使用 Excel 将 X.csv 转成 X.txt 文件（请使用文本文件，而不要使用 Unicode 字符编码的文件），再将文件读入。若文件中有中文，则需先确认文件中的字符编码，转换文件的字符编码后再将文件读入。MacOS 操作系统默认的字符编码是 UTF-8，而 Windows 操作系统的中文版默认的字符编码是 GBK（简体字）或 Big5（繁体字）。使用 Windows 操作系统时，我们可以用 notepad 打开 CSV 文件，再另存为 UTF-8 编码的文件，而后再读入文件。

读取网络上的 iris.csv，地址为 https://github.com/rashida048/Datasets/blob/master/iris.csv：

```
> X <- read.csv("https://raw.githubusercontent.com/uiuc-cse/data-fa14/
gh-pages/data/iris.csv", header = TRUE, encoding = "UTF-8")
> X
```

或读取网络上的 iris.table，地址为 https://archive.ics.uci.edu/ml/machine-learning-databases/iris/iris.data：

```
# 下载iris.dat后，可使用Excel转换文件格式
> X <- read.table("https://archive.ics.uci.edu/ml/machine-learning-
databases/iris/iris.data", sep=",",header = F)
> X
```

R 4.0 以上版本最重要的更新便是导入的字符串数据不再被默认转换成因子变量（Factor），过去 stringsAsFactors 选项默认为 TRUE，因此导入的字符串数据都会被转换成因子变量，但是在新版本中，stringsAsFactors 选项默认为 FALSE。

```
> mode(X$species)
[1] "character"
> class(X$species)
[1] "character"

> X <- read.table("X.txt",header=TRUE, encoding="UTF-8")
> X
  id age     sex   pay
1 1  25      Male  30000
2 2  30      Male  40000
3 3  35      Female 45000
4 4  49      Female 50000
```

scan()函数比 read.table()函数更加灵活，因为 scan()函数可接受键盘输入的数据：

```
> X <- scan("")
1: 12          # 输入值后按Enter键
2: 10
3: 5
4: 6.3
5:             # 不再输入数据时可再按Enter键来结束输入
Read 4 items
> X
[1] 12.0 10.0  5.0  6.3
```

scan()函数也可以指定输入数据的数据类型，例如要创建列表对象：

```
> my=scan(file="",what=list(name="",pay=integer(0),sex=""))
1: peter 50000 M      # 输入值后按Enter键
2: lisa 40000 F
3: johnson 65000 M
4:                    # 不再输入数据时可再按Enter键来结束输入
Read 3 records

> mode(my)
[1] "list"
```

其中：

- file：文件路径，file=""表示由键盘输入值。
- what：设置输入值的数据类型，上述例子为创建列表对象，且其第一个元素name=""表示字符串，第二个元素pay=integer(0)表示整数，而第三个元素sex=""也是字符串。

scan()函数也可以读取 CSV 文件和文本文件，表 2-2 所示为 X1.csv 文件。

表2-2　X1.csv文件

id	age	pay
1	25	30000
2	30	40000
3	35	45000
4	49	50000

```
> X <- scan("X1.csv", sep=",")
Read 12 items
> X
 [1]     1    25 30000     2    30 40000     3    35 45000     4    49 50000
```

使用 Excel 将 X1.csv 转换成 X1.txt 文件再读入：

```
> X <- scan("X1.txt")
Read 12 items
```

```
> X
[1]    1    25 30000    2    30 40000    3    35 45000    4    49    50000
```

2.2 写入数据

若需将存储数据或分析结果输出至外部文件，则可调用 write.table() 函数：

```
> write.table(X,"C:/X_File.csv",row.names=FALSE,col.names=TRUE,sep=",",
+             fileEncoding="GBK")
```

其中：

- X：表示欲输出至外部文件的对象。
- "C:/X_File.csv"：用于指定欲输出至外部文件的文件名和路径。
- row.names：表示输出至外部文件是否加上行名称。
- col.names：表示输出至外部文件是否加上列名称。
- sep=","：表示分隔符。
- fileEncoding：表示输出至外部文件的格式。

R语言提供了一些内建的数据集，可调用data()函数来查询这些内建的数据集，如图2-1所示。

```
> data()
```

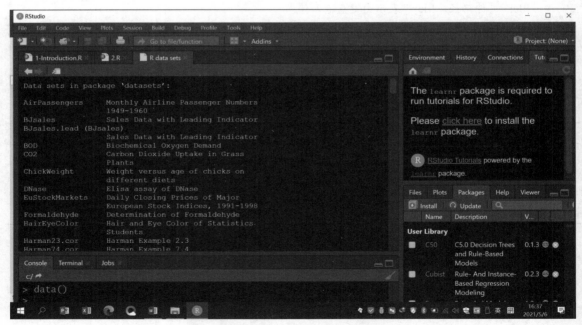

图 2-1　调用 data() 函数查看内建的数据集

可调用 data（数据集名称）函数来使用内建的数据集，例如欲使用 iris 数据集，可执行如下的指令，结果如图 2-2 所示。

使用内建的iris数据集
```
> data(iris)
> iris
```

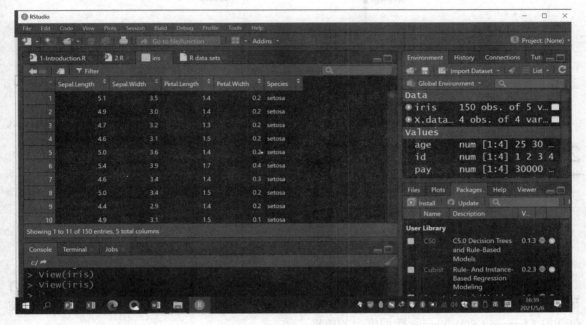

图 2-2　iris 数据集

可调用 str()函数来取得数据集的数据结构：

```
> str(iris)
'data.frame':   150 obs. of  5 variables:
 $ Sepal.Length: num  5.1 4.9 4.7 4.6 5 5.4 4.6 5 4.4 4.9 ...
 $ Sepal.Width : num  3.5 3 3.2 3.1 3.6 3.9 3.4 3.4 2.9 3.1 ...
 $ Petal.Length: num  1.4 1.4 1.3 1.5 1.4 1.7 1.4 1.5 1.4 1.5 ...
 $ Petal.Width : num  0.2 0.2 0.2 0.2 0.2 0.4 0.3 0.2 0.2 0.1 ...
 $ Species     : Factor w/ 3 levels "setosa","versicolor",..: 1 1 1 1 1 1 1 1 1 1 ...
```

2.3　读写 RData 数据

R 语言可将使用的对象存储成 RData 格式的外部文件，也可将 Rdata 格式的外部文件读取回 R 中。若想把 iris 数据集存储至 c:\iris.RData，则可调用 save()函数，如下所示：

```
> setwd("c:/")
> data(iris)
> save(iris,file="iris.RData")
```

若想把 iris.Rdata 文件读取到 R 系统中，则可调用 load()函数，如下所示：

```
> getwd()
[1] "c:/"
```

```
> load("iris.RData", .GlobalEnv)
```

其中：

- .GlobalEnv：表示用户正在使用的工作空间（Workspace）。

2.4 读取 SQL Server 数据库数据

首先安装 RODBC 包，再使用 RODBC 包。

```
> install.packages("RODBC")
> library("RODBC")
```

我们可以先将本书的 IRIS_Data 数据库附加至 MS SQL Server 中（注意需将防火墙端口打开，并在 SQL Server 配置管理器中启用 TCP/IP）。在 Windows 操作系统环境中可使用 RODBC 包读取 Microsoft SQL Server 数据库中的数据。

我们也可以在 CentOS 操作系统环境中使用 RODBC 包读取 Microsoft SQL Server 数据库中的数据。

调用 odbcConnect()函数连接 IRIS_Data 数据库：

```
> db <- odbcConnect(dsn="test", uid="test", pwd="test")
> sqlTables(db)
```

调用 sqlQuery()函数读取 iris 表格（Table）内的数据：

```
> df <- sqlQuery(db, "select * from iris")
```

显示前 6 笔数据：

```
> head(df)
  sepal_length sepal_width petal_length petal_width species
1          5.1         3.5          1.4         0.2  setosa
2          4.9         3.0          1.4         0.2  setosa
3          4.7         3.2          1.3         0.2  setosa
4          4.6         3.1          1.5         0.2  setosa
5          5.0         3.6          1.4         0.2  setosa
6          5.4         3.9          1.7         0.4  setosa

> str(df)
'data.frame':   150 obs. of  5 variables:
 $ sepal_length: num  5.1 4.9 4.7 4.6 5 5.4 4.6 5 4.4 4.9 ...
```

```
$ sepal_width : num  3.5 3 3.2 3.1 3.6 3.9 3.4 3.4 2.9 3.1 ...
$ petal_length: num  1.4 1.4 1.3 1.5 1.4 1.7 1.4 1.5 1.4 1.5 ...
$ petal_width : num  0.2 0.2 0.2 0.2 0.2 0.4 0.3 0.2 0.2 0.1 ...
$ species     : Factor w/ 3 levels "setosa","versicolor",..: 1 1 1 1 1 1 1 1 1 1 ...
```

结束 ODBC。

```
> odbcClose(db)
```

2.5 读写 Excel 数据

安装 xlsx 包，再使用 xlsx 包：

```
> install.packages("xlsx")
> library("xlsx")
```

设置工作目录，再使用 iris 数据集：

```
> setwd("c:/")
> data("iris")
```

调用 write.xlsx()函数执行写入操作：

```
> write.xlsx(iris, file="myexcel.xlsx", sheetName="IRIS", append=T)
> write.xlsx(mtcars, file="myexcel.xlsx", sheetName="MTCARS", append=+TRUE)
> write.xlsx(Titanic, file="myexcel.xlsx", sheetName="TITANIC", appen+d=TRUE)
```

调用 read.xlsx()函数把 myexcel.xlsx 文件的第一个工作表读至 res 变量：

```
# read the first sheet
> sheetIndex=1
> res <- read.xlsx("myexcel.xlsx", sheetIndex, header=TRUE)
```

2.6 习题

从 UCI Machine Learning Repository 下载 iris 数据集，并转成 iris.csv 文件。

第 3 章

从流程控制到函数

R 语言也是一种表达式语言，其所有的指令都是函数或表达式，而赋值表达式"<-"的返回值就是用于赋值的值，在 R 语言中最简单的运行方式就是一行一行地输入表达式，然后显示运行的结果，例如：

```
> a <- c(1,2,3)
> x <- a+2
> x
[1] 3 4 5
```

若欲直接显示运行的结果，则可将指令用括号"()"括起来，例如：

```
> a <- c(1,2,3)
> (x <- a+2)
[1] 3 4 5
```

指令可以用大括号括起来，例如{expr#1;…;expr#m}，以运行多条指令，例如：

```
> {a <- c(1,2,3);x=a+2}
> x
[1] 3 4 5
```

R 语言的流程控制提供条件执行（Condition Execution）与循环（Loop）等结构性语法。

3.1 条件执行

R 语言的条件执行包含 if-else 语句、ifelse()函数和 switch()函数。
R 语言的 if-else 语法为：

```
if (condition) expr#1 else expr#2
```

或

```
if (condition) expr#1
```

其中:

- Condition: 条件判断表达式,必须返回一个布尔值(TRUE 或 FALSE),而&&(AND)和||(OR)逻辑运算符常用于条件判断表达式的条件控制部分。
- expr#1: 一般的表达式。
- expr#2: 一般的表达式。

```
> x <- 6
> if (x>5) y=2 else y=4
> y
[1] 2

> X <- 3
> if (X<5) Y=10
> Y
[1] 10
```

若有多个表达式,则可使用大括号括起来,例如{expr#1;…;expr#m}:

```
> X <- 3
> Y <- 1
> if (X<5 && Y<5)
+ {Y <- 10; Z <- 5}
> Y
[1] 10
> Z
[1] 5
```

R 语言的 ifelse()函数可用于简单的逻辑判断,若 condition 结果为 TRUE,则返回 a;否则返回 b。其语法为:

```
ifelse (condition, a, b)
```

```
> X <- 20
> Y=ifelse(X>5, 2, 3)
> Y
[1] 2
```

R 语言的 switch()函数语法为:

```
switch (condition, expr#1, …, expr#m)
```

其中：

condition 为正整数或文字。若其值为正整数 n，则执行表达式 expr#n，若 n 值大于 m 或小于 1，则 switch() 函数无返回值。若 condition 值为文字，则执行相对应的表达式。

```
> X <- 1
> switch(X, 5, sum(1:10), rnorm(5))
[1] 5
> X <- 2
> switch(X, 5, sum(1:10), rnorm(5))
[1] 55
> X <- 3
> switch(X, 5, sum(1:10), rnorm(5))
[1] -0.185252822 -0.351313575 -0.008195255 -1.920097610 -0.680803488
> X <- 4
> switch(X, 5, sum(1:10), rnorm(5))      # 无返回值
>
> Y <- 1
> switch(Y, juice="Apple", meat="Pork")
[1] "Apple"
```

switch() 函数也可以使用文字，例如：

```
> Y <- "juice"
> switch(Y, juice="Apple", meat="Pork")
[1] "Apple"
```

3.2 循环控制

R 语言的循环控制语句包含 for、while 及 repeat，循环中可使用 break 指令跳出循环，或使用 next 跳过当前一轮循环尚未执行的语句，直接进入当前循环体的下一轮循环。

R 语言的 for 语法为：

`for (index in expr#1) expr#2`

或

`for (index in expr#1) {expr#2;…;expr#m}`

其中：

- index：循环索引。
- expr#1：数值或文字向量，例如 1:5 或 c("A","O","B","AB")。

- expr#2：根据 index 而设计的区块表达式。for 循环会将 expr#1 向量中的每个元素按照顺序以一次一个的方式指定给 index，每指定一次，index 就会运行一次对应的 expr#2 表达式。
- {expr#2;…;expr#m}：多个表达式。

例如：

```
> X <- 0
> for(i in 1:5) X <- X+i
> X
[1] 15
> X <- 0
> Y <- 0
> for(i in 1:5) { X<- X+i; Y <- i^2}
> X
[1] 15
> Y
[1] 25
```

R 语言的 while 循环语法为：

```
while (condition) expr#1
```

或

```
while (condition) {expr#1;…;expr#m}
```

其中：

- condition：当 condition 值为 TRUE 时，运行循环体内的表达式，并重复运行循环体内的指令直到 condition 值为 FALSE 时为止。
- expr#1：一般表达式。
- {expr#1;…;expr#m}：多个表达式。

例如，求 1+2+…+9+10=55。

```
> sum <- 0
> i <- 1
> while (i <= 10) {sum <- sum + i; i <- i + 1} # condition i <= 10
> sum
[1] 55
```

repeat 重复运行表达式，通常在循环中设置检查控制循环的条件并结合 break 指令，break 可用于结束循环，它也是结束 repeat 循环的唯一办法。

R 语言的 repeat 循环语法为：

```
repeat expr
```

其中：

- expr：一个用大括号括起来的区块表达式，必须设置检查循环控制条件，若符合特定的循环控制条件，则利用 break 指令结束循环。

例如，求 1+2+…+9+10=55。

```
> sum <- 0
> i <- 1
> repeat {
+ sum <- sum + i
+ i <- i + 1
+ if ( i > 10 ) break   # 结束循环
+ }
> sum
[1] 55
```

break 指令可用于结束循环，它是结束 repeat 循环的唯一办法；而 next 指令可用来跳过当前一轮循环尚未执行的语句，直接进入下一轮循环。

例如，求 1+3+…+47+49=625。

```
> sum <- 0
> for (i in 1:50)
+ {
+ if ( i %% 2 == 0 ) next    # %%是求偶数
+ sum <- sum + i              # 若i是偶数，则不运行sum <- sum + i
+ }
> sum
[1] 625
```

同其他程序设计语言一样，R 语言也常需要用到循环，但 R 语言循环运行的效率较差，所以应尽量避免使用循环。在 R 语言中，有些函数如 apply()、lapply() 及 sapply() 等，可以更有效率地运行类似循环的指令。

apply(x, MARGIN, FUN, ...)函数的作用是将一个指定函数的计算运用于数组或矩阵的每一列或每一行。

其中：

- x：要进行计算的目标数组或矩阵。
- MARGIN：其值=1 或 2。1 表示行，2 表示列。
- FUN：为指定的函数。

例如，调用 sum() 函数求出数组的每一行的总数。

```
> X <- array(1:24, dim = c(4,6))
> X
     [,1] [,2] [,3] [,4] [,5] [,6]
[1,]    1    5    9   13   17   21
```

```
[2,]    2    6   10   14   18   22
[3,]    3    7   11   15   19   23
[4,]    4    8   12   16   20   24
> apply(X,1,sum)
[1] 66 72 78 84
```

例如，调用 sum()函数求出数组的每一列的总数。

```
> X <- array(1:24, dim = c(4,6))
> X
     [,1] [,2] [,3] [,4] [,5] [,6]
[1,]    1    5    9   13   17   21
[2,]    2    6   10   14   18   22
[3,]    3    7   11   15   19   23
[4,]    4    8   12   16   20   24
> apply(X,2,sum)
[1] 10 26 42 58 74 90
```

lapply(X,FUN,...)函数的作用是将一个指定函数的计算运用于列表对象 X 的每一元素，并返回一个列表对象，且返回的列表对象的长度与原列表对象 X 的长度一致。

例如：

```
> X <- list(a=1:10, b=exp(-1:1))
> lapply(X,sum)
$a
[1] 55

$b
[1] 4.086161
```

sapply(X, FUN, ...)函数的功能与 lapply(X, FUN, ...)函数的功能类似，但其返回一个向量或矩阵对象。

```
> X <- list(a=1:10, b=exp(-1:1))
> sapply(X,sum)
       a        b
55.000000 4.086161
```

3.3 函数

R 语言提供的常用函数可参考附录 C。我们也可以自定义函数，其语法如下：

```
> myfun <- function(arg#1,arg#2,...) {expr#1;…;expr#m}
```

其中：

- arg#1,arg#2,...: 自变量（Argument），自变量可以多于一个。

- expr#1;…;expr#m: 表达式。自定义函数可以不返回函数值，R 语言默认将函数的最后一个表达式的结果作为返回值，也可以调用 return()函数返回函数的值。

例如：

```
> X <- 1
> myfun <- function(X) { Y <- X+2; return (Y) }
> myfun(X)
[1] 3
```

R 语言的自定义函数允许自变量有默认值。若调用函数时没有给自变量传入值，则以默认值作为自变量的传入值；若自变量的传入值与默认值不同，则不使用默认值。

例如：

```
> X <- 2          # 自变量的传入值与默认的值不同
> myfun <- function(X=1) { Y <- X+2; return (Y) }
> myfun(X)
[1] 4
> myfun <- function(X=1) { Y <- X+2; return (Y) }
> myfun()         # 若没有给自变量传入值，则使用默认值X=1
[1] 3
```

在 R 语言的自定义函数中，若欲修改函数外部对象的值，则只有使用"<<-"才能改变自定义函数外部对象的值。

例如：

```
> x <- 1
> myfun <- function(x) { x <- 2; print(x) }
> myfun(x)
[1] 2          # myfun中 x 的值
> x            # 无法改变外部对象 x 的值
[1] 1

> x <- 1
> myfun <- function(x) { x <<- 2; print(x) }
> myfun()
[1] 2          # myfun中改变外部对象 x 的值
> x            # 外部对象 x 的值已改变
[1] 2
```

3.4 习题

（1）使用 while（i <= 9）{ i <- i + 1;sum <- sum + i}完成 1+2+…+9+10=55。
（2）使用 while 和 break 完成 1+2+…+9+10=55。

第 4 章

绘图功能及基本统计

R 语言内建许多绘图函数,这些函数可以显示各种统计图表,还支持开发者自建一些全新的图形。先参考以下 R 语言的绘图示范:

> demo(graphics)
> demo(image)

R 语言的绘图指令(函数)可以分成三个基本的类型:

(1)高级绘图函数(High-level Plotting Functions):创建一个新的图形,可以包括坐标轴及标题等。

(2)低级绘图函数(Low-level Plotting Functions):在一个已经存在的图形上加上其他的图形元素,如额外的点及线等。

(3)交互式绘图函数(Interactive Graphics Functions):允许交互式地用其他设备(如鼠标)在一个现有图形上继续绘制图形信息。

4.1 高级绘图

常用的高级绘图函数如表 4-1 所示。

表4-1 高级绘图函数

函 数	说 明
plot(y)	以索引为横坐标(x 轴)、y 为纵坐标(y 轴)来绘图
plot(x,y)	以 x(x 轴)和 y(y 轴)为坐标来绘图
pie(y)	绘制饼形图
boxplot(y)	绘制盒形图

(续表)

函　　数	说　　明
stem(y)	绘制茎叶图（Stem-and-leaf Plot）
dotchart(y)	绘制点图
hist(y)	绘制直方图
barplot(y)	绘制条形图
contour(x, y, z)	绘制等高线图

高级绘图函数可以通过改变自变量的值来产生不同的绘图效果。plot()函数自变量的设置值如下：

```
plot(x, y,
type = "p",
bty="o",
pch =
lty =
cex =
lwd =
col =
bg =
xlim = NULL, ylim = NULL,
log = "",
main = NULL,
sub = NULL,
xlab = NULL, ylab = NULL,
cex.main =
col.lab =
font.sub =
ann = par("ann"),
axes = TRUE,
...)
```

其中：

- x：x 坐标。
- y：y 坐标。
- type：设置绘图在(x,y)的显示方式。

    ◆ type="p"：画点。
    ◆ type="l"：画线。
    ◆ type="b"：画点，同时在点与点间画线连接。
    ◆ type="s"：阶梯函数（Step Function），为左连续函数。
    ◆ type="S"：阶梯函数，为右连续函数。
    ◆ type="o"：画线，同时穿过画点。
    ◆ type="h"：从点到 x 横轴画垂直线。

- type="n"：不画任何点与线，但容许画坐标轴且建立坐标系统，以用于后续用低级图形函数作图。
- bty：设置图形坐标轴外框（Box）的类型，其选项共有 { "o", "l", "7", "c", "u", "]" }。
- pch：画点时，设置 pch=k，k 是一个介于 1～25 的正整数，显示一个对应 k 的特定符号，其默认值为 1。
- lty：画线时，设置线条类型。
  - lty=1：实线。
  - lty=2：虚线。
- lwd：画线时，设置线条宽度。
  - lwd=1：默认值。
  - lwd=k：线条宽度的倍数。
- col：设置点、线的颜色。默认值为 black，设置颜色值（Value）为调色板的数值，或者颜色的文字名称。
- bg：设置图形背景颜色，默认值为白色，例如 bg="white"。
- xlim：设置坐标轴 x 值的上下界，如 xlim=c(1, 10)。
- ylim：设置坐标轴 y 值的上下界，如 ylim=c(1, 10)。
- log：设置坐标轴是否取对数值。
  - log="x"：对 x 坐标轴取对数值。
  - log="y"：对 y 坐标轴取对数值。
  - log="xy"：对 x 与 y 坐标轴都取对数值。
- main：设置图形主标题（Main Title），定义的文字放在图形的上方，例如 main="Title"。
- sub：设置图形次标题（Subtitle），定义的文字放在图形的下方，例如 sub="Subtitle"。
- xlab：设置 x 坐标轴的标记，例如 xlab="X"。
- ylab：设置 y 坐标轴的标记，例如 ylab="Y"。
- ann：是否画出自动设置的主标题及坐标轴标记，ann=TRUE 时会画出自动设置的主标题及坐标轴标记。
- axes：是否画出自动设置的坐标轴与坐标轴外框，axes=TRUE 时会画出自动设置的坐标轴与坐标轴外框。

例如：

```
> x <- sin(1:20)
> plot(x, type="l",main="Sin Plot", xlab="X",ylab="Y")
```

结果如图 4-1 所示。

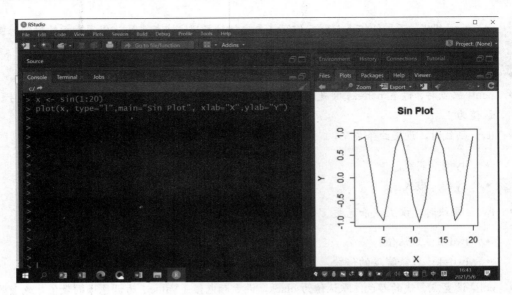

图 4-1　sin 图

4.2　低级绘图

常用的低级绘图函数如表 4-2 所示。

表4-2　低级绘图函数

函　　数	说　　明
points(x,y)	在现有的图形上加一个点
lines(x,y)	在现有的图形上加一条线
text(x,y,labels=z.vec,...)	在(x,y)坐标点标出由 labels 设置的对应 z.vec 的数值型或文字型向量
abline(a,b)	在现有的图形上加画一条截距为 a 和斜率为 b 的直线
abline(h=y)	画出满足 Y=y 且平行于 x 坐标轴的水平线
abline(v=x)	画出满足 X=x 且垂直于 x 坐标轴的垂直线
polygon(x,y,...)	画出以(x,y)坐标点为顶点的多边形（Polygon），可以用 col=自变量指定一个特定颜色来填满多边形的内部
legend(x,y,leg.vec,...)	在现有图形的(x,y)坐标位置绘制图例（Legend），图例的说明文字由向量 leg.vec 表示
title(main,sub)	main="My Main Title"设置图形主标题，定义的文字放在图形的上方，sub="My Subtitle"用于设置图形的次标题，定义的文字放在图形的下方
mtext(text,side=3,line=1)	在现有图形的边缘，加上文字

例如：

```
> x <- sin(1:20)
> plot(x, type="l", xlab="X",ylab="Y")
> title(main="Sin Plot",sub="图4-2:低级绘图函数图")
```

结果如图 4-2 所示。

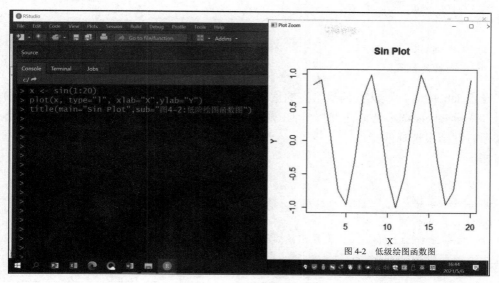

图 4-2　低级绘图函数图

4.3　交互式绘图

R 语言同时提供了允许我们直接用鼠标在一个图形上抽取（Extract）和增加（Add）信息的函数，其中常用的是 locator() 和 identify() 函数。

locator() 函数让我们可以通过鼠标左键点选当前图形上的特定位置：

```
locator(n,type)
```

其中：

- n：指定要点选几个坐标点，若不指定，则默认 n=512。
- type：允许在被选择的点上画图，并且有高级绘图函数一样的效果；默认情况下不能画图。
locator() 函数使用具有 x 和 y 形式的列表对象返回所选中点的位置信息。

例如：

```
> plot(2, 2)
> pts <- locator(n = 3)      # 可在图形中选择三个坐标点
> pts                        # 选择完成后，pts的值
$x
[1] 1.621502 1.840632 2.395573

$y
[1] 1.771951 2.347735 1.522875
```

identify() 函数容许我们将定义的标签（Label）用鼠标左键放置在鼠标指针的选择处。

```
identify(x,y,labels)
```

其中：

- x：x 坐标的位置。
- y：y 坐标的位置。

在设置 labels 时，默认值为显示点的索引，选择完成后，单击鼠标右键结束选择的操作；当使用自变量 labels="My Labels"时，可在鼠标左键点选处显示我们定义的标签 My Labels，并单击鼠标右键结束选择的操作。

例如：

```
> x <- c(1, 3, 5, 7, 8, 9, 3, 6, 7, 2)
> y <- c(5, 3, 5, 8, 2, 1, 4, 3, 4, 7)
> plot(x, y)
> sel <- identify(x, y) # 单击鼠标左键10次
```

结果如图 4-3 所示。

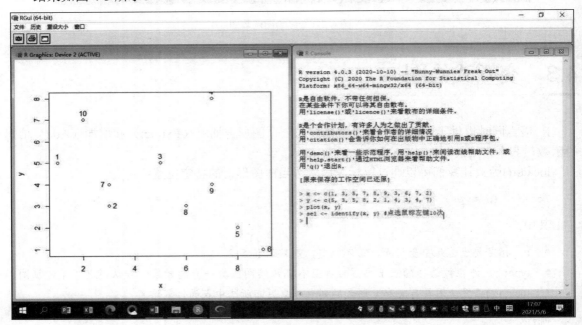

图 4-3　identify()函数运行后的结果图

在选择的过程中 identify()函数会在选中的数据点旁边标上我们所定义的标签，而这些标签位置的索引会在选择完成后返回给 sel 变量。最后我们就可以靠 sel 变量取得选择的数据点：

```
> x <- c(1, 3, 5, 7, 8, 9, 3, 6, 7, 2)
> y <- c(5, 3, 5, 8, 2, 1, 4, 3, 4, 7)
> plot(x, y)
> sel <- identify(x, y,"MY LBAELS")    # 单击鼠标左键选择图中间的位置
                                        # 然后单击鼠标右键结束选择
> x.sel <- x[sel]
```

```
> y.sel <- y[sel]
> x.sel
[1] 5
> y.sel
[1] 5
```

结果如图 4-4 所示。

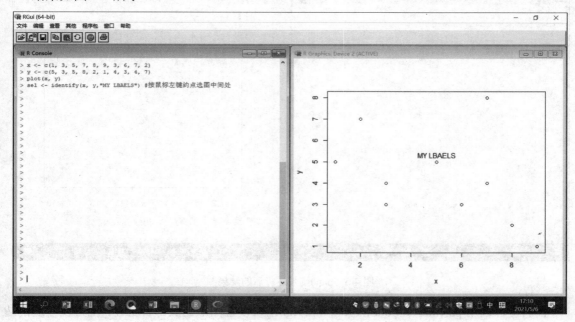

图 4-4　identify(x,y,"MY LBAELS")运行后的结果图

4.4　图形参数

R 语言提供了许多图形参数（Graphics Parameters）用以控制图形的颜色、文字对齐等。直接调用 par()函数可以得到当前所有这些参数的设置值，例如：

```
> par()
```

通过调用 par()函数也可以更改或设置这些图形参数：

```
par(par.name=par.value)
```

其中：

- par.name：图形参数名称，可用于 par()函数或某些（高级或低级）绘图函数中作为自变量。
- par.value：par.name 图形参数的设置值。

例如：

```
> x <- c(5, 3, 5, 8, 2, 1, 4, 3, 4, 7)
```

```
> par(col=4, lty=4)    # 设置参数
> plot(x, type="l", xlab="X",ylab="Y")
```

结果如图 4-5 所示。

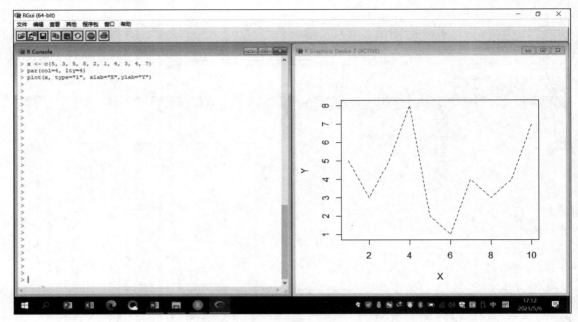

图 4-5　par()函数运行后的结果图

我们可以调用 par(mar=c(bottom, left, top, right))来设置图形到底部、左边、上方及右边的边距，单位为 cm；也可以调用 par(mfrow = c(nr, nc))显示 nr×nc 个子图。

例如：

```
par(mfrow=c(1,2))
par(mar=c(5, 4, 4, 2))
par(col=4, lty=1)

plot(x, type="l", xlab="X",ylab="Y")
barplot(x, xlab="X",ylab="Y")
```

结果如图 4-6 所示。

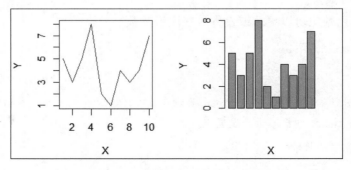

图 4-6　调用 par()函数画 1×2 子图

4.5 基本统计

描述统计学（Descriptive Statistics）的主要目的是通过对数据样本进行综合和概括，并通过图形呈现出数据样本的特性，从而了解数据的规律性特征。R 语言提供了 summary()函数来取得数据及其分布的信息。

```
> summary(iris)
  Sepal.Length    Sepal.Width     Petal.Length    Petal.Width          Species
 Min.   :4.300   Min.   :2.000   Min.   :1.000   Min.   :0.100   setosa    :50
 1st Qu.:5.100   1st Qu.:2.800   1st Qu.:1.600   1st Qu.:0.300   versicolor:50
 Median :5.800   Median :3.000   Median :4.350   Median :1.300   virginica :50
 Mean   :5.843   Mean   :3.057   Mean   :3.758   Mean   :1.199
 3rd Qu.:6.400   3rd Qu.:3.300   3rd Qu.:5.100   3rd Qu.:1.800
 Max.   :7.900   Max.   :4.400   Max.   :6.900   Max.   :2.500
```

常用的统计图为直方图及盒形图。直方图又称柱状图，是用来表现单变量数据最常见的图表，可呈现数据的分布状况。

例如：

```
> y=c(170,170,171,172)
> hist(y,col='grey')   # 直方图
```

结果如图 4-7 所示。

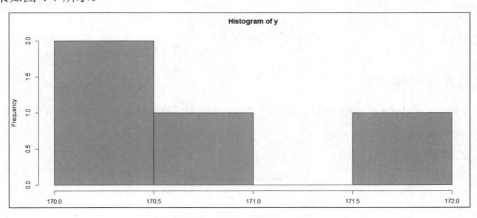

图 4-7　直方图

盒形图又称箱形图或盒须图，可显示出数据的最大值、最小值、中位数、第一个四分位数及第三个四分位数。若将数据从小到大排列并分成四等份，则会有三个分割点。从最小的那一边算起，第一个分割点称为第一个四分位数；第二个分割点称为第二个四分位数，也就是中位数；第三个分割点称为第三个四分位数。

例如：

```
> y1=c(165,166,167,167,175,176,177,178,179,180)
```

```
> median(y1,na.rm=TRUE)        # 中位数
[1] 175.5
```

其中:

- na.rm=TRUE: 表示若有 NA 值, 则删除。

```
> max(y1)                      # 最大值
[1] 180

> min(y1)                      # 最小值
[1] 165

> max(y1)-min(y1)              # 全距
[1] 15

> range(y1)
[1] 165 180

> quantile(y1,0.25)            # 第一个四分位数
25%
167

> quantile(y1,0.75)            # 第三个四分位数
  75%
177.75

> IQR(y1)                      # 四分位数间距
[1] 10.75
```

在垂直方向画盒形图:

```
> boxplot(as.data.frame(y1), main = "boxplot(*, horizontal = FALSE)", horizontal = FALSE)
```

结果如图 4-8 所示。

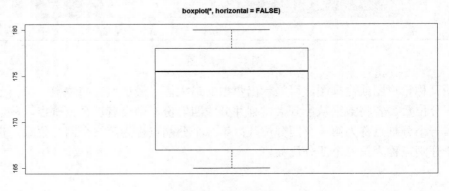

图 4-8 盒形图

常用的描述性统计函数有均值（Mean，即算术平均值）、中位数（Median）、众数（Mode）、方差（Variance）、标准差（Standard Deviation，也称为标准偏差）及相关系数（Correlation）。

均值公式为：

$$\bar{x} = \frac{\sum_{i=1}^{N} x_i}{N} \tag{4-1}$$

中位数为数据经过排序后的中间值。若样本数为偶数，则取中间两个数的平均值。众数是指一组数据中出现次数最多的那个数。

方差公式为：

$$var_x = \frac{\sum_{i=1}^{N}(x_i - \bar{x})^2}{N-1} \tag{4-2}$$

标准差公式为：

$$S_x = \sqrt{var_x} \tag{4-3}$$

相关系数公式为：

$$r_{xy} = \frac{\sum_{i=1}^{N}(x_i - \bar{x})(y_i - \bar{y})}{(N-1)S_x S_y} \tag{4-4}$$

例如：

```
> y1=c(165,166,167,167,175,176,177,178,179,180)
> median(y1,na.rm = TRUE)      # 中位数
[1] 175.5

> var(y1)                      # 方差
[1] 36

> sd(y1)                       # 标准差
[1] 6

> table(y1)                    # 出现次数
y1
165 166 167 175 176 177 178 179 180
  1   1   2   1   1   1   1   1   1
> which.max(table(y1))         # 众数及其排列位置
167
  3
> cor(y1,y1)                   # 相关系数
[1] 1

> cor(y1,-y1)
[1] -1
```

回归（Regression）分析是一种统计分析方法，目的在于了解两个或多个变量间是否相关，观察特定变量来预测我们感兴趣的变量，并建立因变量（即函数值）与自变量（或称解释变量）间关系的数学模型。线性回归是利用一个含有单一或多个自变量的回归公式来预测因变量，如下所示：

$$y = c_0 + c_1 x_1 + c_2 x_2 + \ldots + c_k x_k \tag{4-5}$$

其中：

- y：因变量。
- c_i：回归系数。
- x_i：自变量。
- c_0：截距。

例如：

```
> setwd("D:/")                # 设置工作目录
> A10 <- read.table(file="grade.csv",header=TRUE,sep=",",encoding="GBK")
> str(A10)
'data.frame':   10 obs. of  3 variables:
 $ 学生编号      : int  1 2 3 4 5 6 7 8 9 10
 $ 每周自修时数.X : num  10.5 9.2 11.6 6.3 8.2 12.1 15.2 8.5 7.5 10.2
 $ 考试成绩.Y    : int  91 86 89 81 84 92 96 83 77 87
> A10 <- as.matrix(A10)       # 将A10对象由数据框转成矩阵并删除header
> A10 <- matrix(A10, ncol = ncol(A10), dimnames = NULL)
> X=A10[,2]
> Y=A10[,3]
> Lm_model <- lm(Y ~ X)       # 运行回归
> Lm_model

Call:
lm(formula = Y ~ X)

Coefficients:
(Intercept)            X
     66.579        2.016
```

其回归公式可表示为：

$$Y = 66.579 + 2.016 X \tag{4-6}$$

也可以调用 coef() 来取得截距及回归系数：

例如：

```
> cf <- coef(lm(Y ~ X))       # 取得截距及回归系数
> cf
(Intercept)           X
  66.579002    2.016213
```

若要验证回归公式的输出值，则可自定义函数及调用 sapply()函数：

```
> lm_function <- function(x) {y <- cf[1]+cf[2]*x; return (y) }
> Y_output <- sapply(X,lm_function)
> Y_output
 (Intercept) (Intercept) (Intercept) (Intercept) (Intercept) (Intercept)
(Intercept)
    87.74924    85.12816    89.96708    79.28115    83.11195    90.97518
97.22544
 (Intercept) (Intercept) (Intercept)
    83.71682    81.70060    87.14438
```

可调用 abs()函数计算 Y-Y_output 的绝对值：

```
> abs(Y-Y_output)
 (Intercept) (Intercept) (Intercept) (Intercept) (Intercept) (Intercept)
   3.2507584   0.8718357   0.9670761   1.7188541   0.8880489   1.0248172
 (Intercept) (Intercept) (Intercept) (Intercept)
   1.2254439   0.7168150   4.7006018   0.1443776
```

调用 par()函数设置画图参数，调用 plot()函数和 abline()函数画出回归公式，并调用 points()函数在图形上标出 Y_outpot：

```
> par(mfrow=c(1,1))
> par(mar=c(5,4,4,2))
> par(col="black")
> plot(X,Y)
> abline(lm(Y ~ X))
> par(col="blue")
> points(X,Y_output)
```

结果如图 4-9 所示。

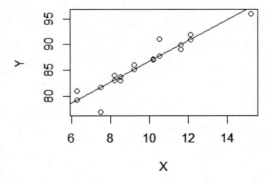

图 4-9　回归公式对应的图形

4.6 习题

bank.csv 数据集的分析目标是预测客户认购银行的定期存款产品的影响因素。数据的实例数：4521，属性数：12。文档是关于银行定期存款预测情况，采用 Y（定期存款产品的订购）作为目标变量，采用 age（客户年龄）、job（工作类型）、martial（婚姻状况）、education（教育程度）、default（信用违约）、housing（住房贷款）、loan（个人贷款）、poutcome（营销活动）作为解释变量。

（1）调用 summary() 函数提供最小值、最大值、四分位数和数值型变量的均值，以及因子向量和逻辑型向量的频数统计。

（2）订购银行定期存款在 age 方面的 Histogram（直方图）分布。

（3）订购银行定期存款在 job、marital、education、default、housing、loan、poutcome 等方面的 Bar Plot（条形图）分布。

第 5 章

数据分析和常用的包介绍

数据分析是一个跨领域的交叉学科,通过数据采集、数据清理、建立算法、可视化来发现知识,进而提出合理的预测和建议。数据分析是多个领域之间的协作行为,本章介绍 R 语言在机器学习(Machine Learning)、数据挖掘(Data Mining)及文本挖掘(Text Mining)方面的常用包。

5.1 机器学习介绍

机器学习就是让机器(计算机)具有学习能力及从数据中自动学习规则,并利用规则对新的数据进行预测,其主要是设计和分析可以自动学习的算法,让计算机可以从过去的数据或经验中建立一个模型(Model),而学习(Learning)就是运行此模型,并利用训练数据集(Training Dataset)来建立模型。

常用的机器学习可以分成以下几种类别:

(1)监督式学习(Supervised Learning):可以从训练数据中学到或建立一个模型,并依此模型预测新的案例。训练数据是由输入数据和预期输出组成的。分类(Classification)就是常见的监督式学习算法。在机器学习领域中,可结合多个分类模型,以达到更佳的分类性能,而这种方法称为集成学习方法(Ensemble Methods,或称为组合学习方法)。

(2)非监督式学习(Unsupervised Learning):与监督式学习不同的是,训练数据中并无预期输出。聚类(Clustering)就是常见的非监督式学习算法。

(3)演化式学习(Evolutionary Learning):主要是基于模仿生物演化及行为而发展出来的学习算法。遗传算法(Genetic Algorithm)就是典型的演化式学习算法。

(4)混合式学习(Hybrid Learning): 主要是结合多种学习法的优点,借以提升学习的性能(Performance)或效率(Efficiency)。

5.2 数据挖掘介绍

Fayyad（1996）曾将数据库知识发现（Knowledge Discovery in Database，KDD）的过程定义为："从数据中建立确定有效的、新颖的、潜在有用的以及易于理解的模型的过程，且此过程不是显而易见的"。数据挖掘是整个知识发现中的核心步骤，Berry & Linoff（1997）将其定义为："为了发现有意义的模型或规则，必须从大量数据中以自动或半自动的方式来探索和分析数据"。

常用的数据挖掘方法可以分成以下几种类别：

（1）分类：将数据中各属性（Attribute）分门别类地加以定义，通过训练大量数据后得到的规则来建立类别（Class）模型。

（2）聚类：按定义的相似程度将数据分为不同的簇（Cluster）。其中相似程度可以利用不同的距离或相似度（Similarity）来定义。聚类与分类最大的不同在于，聚类并没有预先定义好类别，而聚类得到的簇的意义要靠分析者事后的阐释才能确定。

（3）关联规则（Association Rule）：关联规则的目的是找出数据间可能相关的项目，通过数据寻找同时发生的事件（Event）或记录（Record），借以推导出其间的关联规则。

R 语言提供了许多与机器学习及数据挖掘有关的包。常用于分类的监督式学习算法包含决策树（Decision Tree）、支持向量机（Support Vector Machine）、人工神经网络（Artificial Neural Network）及集成学习方法（Ensemble Method），其对应的包为 rpart、C50、e1071、neuralnet、randomForest 及 adabag。常用于聚类的非监督式学习算法包含 K 均值聚类算法（K Means）及模糊 C 均值聚类算法（Fuzzy C Means），其对应的包为 e1071。演化式学习有许多不同的算法，其中包含遗传算法及人工蜂群（Artificial Bee Colony）算法，其对应的包为 GA 和 ABCoptim。关联规则分析对应的包为 arules。

5.3 文本挖掘介绍

文本挖掘的特点在于它的原始输入数据都是没有特定结构的纯文本，这些文本的内容是由人类用自然语言所写成的，所以无法直接使用数据挖掘的算法来探索和分析数据。R 语言提供的文本挖掘的包有 gutenbergr 和 jiebaR。

5.4 常用的包介绍

本节将介绍常用的包：rpart、C50、e1071、neuralnet、randomForest、adabag、GA、ABCoptim、arules、Rfacebook、wordcloud、jiebaR、gutenbergr、rmr2、spark.kmeans、spark.mlp 及 spark.randomForest。

rpart 包是由 Breiman、Friedman、Olshen 和 Stone 等人所撰写的图书 *Classification and Regression Trees*（1984）中提出的重要方法。rpart 包中的 rpart()函数的基本语法及其重要自变量如下：

```
rpart(formula, data, weights, subset, na.action=na.rpart, method, parms, control,...)
```

其中：

- formula：模型的公式，例如 Y~X1+X2+...+XK。
- data：建立模型用的数据。
- weights：data 的权重，为非必要的自变量。
- subset：只使用部分数据（子集合），为非必要的自变量。
- na.action：对缺失值的处理方式，默认会删除。
- method：使用的方法，例如 anova、poisson、class 及 exp。
- parms：会根据不同的方法给予不同的自变量，若为 anova 方法，则不需要设置此自变量。
- control：控制此算法的自变量，可在 rpart.control 中设置。

rpart.control()函数的基本语法及其重要自变量如下：

```
rpart.control(minsplit=20, minbucket=round(minsplit/3), cp =0, maxdepth=30,...)
```

其中：

- minsplit：建立一个新节点（Node）时最少需要几笔数据。
- minbucket：建立叶节点（Leaf Node）时最少需要几笔数据。
- cp：决定计算复杂度（Complexity）的参数，用于修剪树的分支。
- maxdepth：树的深度。

rpart()函数建立模型后可调用 predict()函数来预测此模型的结果。predict()函数的基本语法及其重要自变量如下：

```
predict(object, newdata, type, na.action,...)
```

其中：

- object：用来进行预测的模型。
- newdata：测试数据集。
- type：使用的学习方法，例如 type="class"是使用分类。
- na.action：对缺失值的处理方式，默认不会删除（注意：与 rpart()函数处理方式不同）。

C50 包是由 Quinlan 提出的 C5.0 决策树。C50 包中的 C5.0()函数的基本语法及其重要自变量如下：

```
C5.0(x, y, trials=1, rules=FALSE, weights=NULL, control= C5.0Control(),...)
```

其中：

- x: 输入属性或自变量。
- y: 输出属性或因变量（即解释变量）。
- trials: boosting 的迭代（Iteration）次数，trials=1 表示只使用一种模型。
- rules: 是否输出为规则，而非树状结构。
- weights: 数据的权重。
- control: 控制此算法的自变量，可在 C5.0 Control()中设置。

C5.0 Control()函数的基本语法及其重要自变量如下：

```
C5.0Control(subset=TRUE,winnow=FALSE,noGlobalPruning=FALSE,CF,minCases,sample,seed,label="outcome")
```

其中：

- subset: 是否使用部分数据（子集合）。
- winnow: 是否使用属性筛选。
- noGlobalPruning: 是否运行决策树修剪。
- CF: 为信赖水平，其值介于(0,1)之间。
- minCases: 建立一个节点时最少需要几笔数据（案例）。
- sample: 用于训练数据的比例，其值介于(0,0.999)之间。
- seed: 随机数。
- label: 输出属性的标签。

C5.0()函数建立模型后可调用 predict()函数来预测此模型的结果。predict()函数的基本语法及其重要自变量如下：

```
predict(object, newdata, trials, type, na.action,...)
```

其中：

- object: 用来做预测的模型。
- newdata: 测试数据集。
- trials: 作为预测时 boosting 迭代的次数。
- type: 使用 class 或 prob，例如 type="class"是使用分类。
- na.action: 对缺失值的处理方式。

e1071 包中包含支持向量机学习法。e10710 包中的 svm()函数的基本语法及其重要自变量如下：

```
svm(formula, data, type, kernel, gamma, cost, subset, na.action, scale=TRUE,...)
```

其中：

- formula: 模型的公式，例如 Y~X1+X2+...+XK。
- data: 建立模型用的数据。
- type: 使用的学习方法，包含 C-classification nu-classification、one-classification、eps-regression、nu-regression。

- kernel：核（Kernel）函数，包含 linear、polynomial、radial base 及 sigmoid 函数。
- gamma：除了核函数为 linear 外，所有核函数使用的 γ 值。
- cost：C 的值。
- cross：若 K>0，则使用 K-fold。
- subset：使用部分数据（子集合）。
- na.action：对缺失值的处理方式。
- scale：属性数据是否要正规化。

svm()函数建立模型后可调用 predict()函数来预测此模型的结果。predict()函数的基本语法及其重要自变量如下：

```
predict.svm(object, newdata, na.action,...)
```

其中：

- object：用来做预测的模型。
- newdata：测试数据集。
- na.action：对缺失值的处理方式。

e1071 包提供了 tune.svm()函数用于搜索最佳 gamma(γ)及 cost(C)值。tune.svm()函数的基本语法及其重要自变量如下：

```
tune.svm(formula, data, gamma, cost,...)
```

其中：

- formula：模型的公式。
- data：建立模型用的数据。
- gamma：指定搜索γ值的范围。
- cost：指定搜索C值的范围。

e1071 包提供了运行模糊 C 均值聚类算法的 cmeans()函数。cmeans()函数的基本语法及其重要自变量如下：

```
cmeans(x, centers, iter.max, verbose, dist, method, m=2,...)
```

其中：

- x：作为聚类的数据。
- centers：聚类的簇数（Cluster Number）。
- iter.max：最大的迭代次数。
- verbose：设为 TRUE 时，可显示运行过程的信息。
- dist：计算距离的公式，例如 euclidean 和 manhattan。
- method：使用何种学习方法，例如 cmeans。
- m：模糊 C 均值聚类算法的参数值。

neuralnet 包提供了人工神经网络学习法。neuralnet 包中的 neuralnet()函数的基本语法及其重要自变量如下：

```
neuralnet(formula, data, hidden, threshold, stepmax, rep, startweights,
learningrate.limit, learningrate, algorithm,...)
```

其中：

- formula：模型的公式。
- data：建立模型用的数据。
- hidden：隐藏层神经元（Neuron）的数量（值得注意的是，此函数只提供 1 层隐藏层）。
- threshold：阈值。
- stepmax：训练时的最大迭代次数。
- rep：训练次数。
- startweights：开始时的权重值。
- learningrate.limit：学习率（Learning Rate）的最大值及最小值。
- learningrate：当使用反向传播（Backpropogation）算法时的学习率。
- algorithm：使用的算法，例如 backprop、rprop+、rprop-、sag 或 slr。

neuralnet 包提供了 compute() 函数来预测此模型的结果，compute() 函数的基本语法及其重要自变量如下：

```
compute(x, covariate, rep)
```

其中：

- x：用来做预测的模型。
- covariate：测试数据集。
- rep：预测次数。

randomForest 包提供了集成学习方法的随机森林学习法。randomForest 包中的 randomForest() 函数的基本语法及其重要自变量如下：

```
randomForest(formula, data, ntree, na.action,...)
```

其中：

- formula：模型的公式。
- data：建立模型用的数据。
- ntree：决策树的数量。
- na.action：数据中有缺失值时所调用的函数。

randomFores 包提供了 predict() 函数来预测此模型的结果，predict() 函数的基本语法及其重要自变量如下：

```
predict(object, newdata,...)
```

其中：

- object：用来做预测的模型。
- newdata：测试数据集。

adabag 包提供了集成学习方法的提升法（Boosting）。adabag 包中的 boosting() 函数的基本语法及其重要自变量如下：

```
boosting(formula, data, boos = TRUE, mfinal = 100,...)
```

其中：

- formula：模型的公式。
- data：建立模型用的数据。
- boos：若设为 TRUE（默认值），则权重以该次迭代运算所使用的训练集的观察值重新计算；若为 FALSE，则使用相同的权重。
- mfinal：迭代运算的次数，默认值为 mfinal=100（整数）。

adabag 包提供了 predict()函数来预测此模型的结果，predict()函数的基本语法及其重要自变量如下：

```
predict(object, newdata, newmfinal,...)
```

其中：

- object：用来做预测的模型。
- newdata：测试数据集。
- newmfinal：进行预测时的迭代次数。

NbClust 包提供了 NbClust()函数来获得聚类指标（Clustering Index），借以评估聚类的效果。NbClust()函数的基本语法及其重要自变量如下：

```
NbClust (data, distance = "euclidean",min.nc=2, max.nc=15, method = "kmeans",
index = "all",...)
```

其中：

- data：用于聚类的数据。
- distance：计算距离的公式，例如 euclidean 和 manhattan。
- min.nc：设置最小的簇数量。
- max.nc：设置最大的簇数量。
- distance：计算距离的公式，例如 euclidean 和 manhattan。
- method：使用何种学习方法，例如 kmeans。
- index：聚类指标，例如 kl、ch、hartigan、ccc、scott、marriot、trcovw、tracew、friedman、rubin、cindex、db、silhouette、duda、pseudot2、beale、ratkowsky、ball、ptbiserial、gap、frey、mcclain、gamma、gplus、tau、dunn、hubert、sdindex、dindex、sdbw，若为 all，则使用除了 gap、gamma、gplus 及 tau 之外的所有聚类指标。

GA 包提供了 ga()函数运行遗传算法。ga()函数的基本语法及其重要自变量如下：

```
ga(type, fitness, min, max, nBits, population, selection, crossover, mutation,
popSize, pcrossover, pmutation, elitism, monitor, maxiter, run, maxfitness,...)
```

其中：

- type：运行遗传算法的方式，例如 binary、real-valued 及 permutation。
- fitness：适应性函数（Fitness Function）值。

- min: 搜索空间（Search Space）的最小值。
- max: 搜索空间（Search Space）的最大值。
- nBits: type="binary"时的位数。
- population: 随机产生的初始族群（Population）。
- selection: 遗传算法的选择（Selection）机制。
- crossover: 遗传算法的交配（Crossover）机制。
- mutation: 遗传算法的突变（Mutation）机制。
- popSize: 族群规模（Size）。
- pcrossover: 产生交配的概率。
- pmutation: 发生突变的概率。
- elitism: 多少百分比的个体（Individual）会保留至下一代（Generation）。
- monitor: 显示运行的过程。
- maxiter: 为最大迭代次数。
- run: 若连续几代皆无法改善最佳值，则停止运行遗传算法。
- maxfitness: 运行遗传算法后找到的最大适应函数值。

ABCoptim 包提供了 abc_optim()函数运行人工蜂群算法。abc_optim()函数的基本语法及其重要自变量如下：

```
abc_optim(par, fn, D, NP, FoodNumber,lb=-Inf, ub=+Inf, maxCycle=1000, criter=50,...)
```

其中：

- par: 解的初始值。
- fn: 求最小值的目标函数。
- D: 解的参数（Parameter）数量。
- NP: 人工蜂群的蜜蜂数量。
- FoodNumber: 蜜蜂欲搜索的食物数量。
- lb: 搜索空间的下界（Lower Bound）。
- ub: 搜索空间的上界（Upper Bound）。
- maxCycle: 为最大迭代次数。
- criter: 结束条件。

arules 包提供了 apriori()函数运行关联分析算法。apriori()函数的基本语法及其重要自变量如下：

```
apriori(data, parameter = NULL, appearance = NULL, control = NULL)
```

其中：

- data: 可以强制转换为事务历史记录的对象。
- parameter: 自变量，默认自变量值为支持度（Support）=0.1、置信度（Confidence）=0.8，最多规则数量=10。

- appearance: 使用此自变量可以限制显示项目。默认情况下，所有项目都可以显示不受限制。
- control: 控制此算法的自变量。

使用 wordcloud 包可产生词云，wordcloud()函数的基本语法及其重要自变量如下：

```
wordcloud(words,freq,min.freq=2,max.words=Inf,random.order=TRUE,...)
```

其中：

- words: 要显示的词。
- freq: 要显示的词出现的频率。
- min.freq: 要显示的词出现的最小频率。
- max.words: 要显示的词出现的最大频率。
- random.order: 若设置为 TRUE，则显示词的顺序会随机出现；若设置为 FALSE，则显示的词会按频率递减出现。

Project Gutenberg（PG）是把公版著作（Public Domain）数字化制作成电子书，放在网络（https://www.gutenberg.org/）上供用户自由取用。可调用 gutenbergr 包中的 gutenberg_download()函数下载 Project Gutenberg 中数字化的文化作品或电子书：

```
gutenberg_download(gutenberg_id, strip,…)
```

其中：

- gutenberg_id: 要下载的 Project Gutenberg ID，可以是向量或数据框对象。
- strip: 设置为 TRUE 时可去除 Header 和 Footer 内容。

在进行文本挖掘时，中文与英文的处理方式有很大不同，英文有空白作为词与词的自然分词工具，而中文的词与词则是连接在一起的，所以在进行中文文本的挖掘前，需要对文章或句子进行分词的处理，也就是要将词与词分开，以便进行进一步的分析。可调用 jiebaR 包中的 worker()函数来初始化分词类型，再调用 segment()函数进行分词，work()的基本语法及其重要自变量如下：

```
worker(type,dict,...)
```

其中：

- type: 分词类型，其类型有 mix（混合）、mp（最大概率）、hmm、query（索引）、tag（标记）、simhash 及 keywords（关键词）。默认的类型为混合模式。
- dict: 主要词典（Main Dictionary）的路径。

```
segment(code, jiebar,…)
```

其中：

- code: 中文的文章或句子。
- jiebar: 使用 jiebaR Worker。

第 6 章

监督式学习

本章介绍常用的监督式学习算法及其应用,例如决策树、支持向量机、人工神经网络及集成学习方法。

6.1 决策树

在机器学习的分类方法中,决策树可以说是最具代表性的方法,其具有不错的预测准确率(Predict Accuracy)与可解读性(Interpretability),使得决策树广为各个领域使用。决策树在建立过程中会创建一个树结构,该结构由根节点(Root Node)、内部节点(Internal Node)、叶节点(Leaf Node)(或称为类别(Class))组成。决策树停止再往下生长的情况是:该分类数据中的每一笔数据都已经归类到相应的类别下,该类数据中已经没有办法再找到新的属性来进行节点分割或该类数据中已经没有任何尚未处理的数据。在建立树结构后,测试新数据时是从决策树根部节点开始进行测试的,根据各划分选择属性值,移至另一个内部节点,依此以递归方式继续进行,直至到达叶节点,则此叶节点就是一个预测的类别。从根节点自上向下经由内部节点最后到达叶节点,即为一个预测分类结果的规则。决策树的树结构图如图 6-1 所示。

当原始训练数据属性太多或由于决策树算法属性选择有偏好时,就容易因为过度学习而出现决策树过拟合(Over-fitting)的问题,使得所产生的树结构太过于复杂,因此必须进行适当的剪枝。剪枝的方式可分为前剪枝(Pre-Pruning)和修剪枝(Post-Pruning)。前剪枝是运用统计方法来评估是应该继续分割某内部节点还是应该立刻停止。后剪枝允许决策树过拟合的情况存在,在决策树建立完成后再进行剪枝。

目前常见的决策树算法包括分类与回归树(Classification and Regression Trees,CART)、ID3(Inductive Dichotomiser 3)及 C5.0 等。

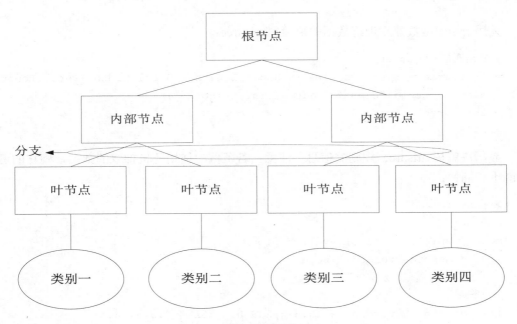

图 6-1　决策树的树结构图

分类与回归树是 Breiman 等学者开发出来的算法，其运算方式主要是利用二分法（Binary）。在分类与回归树运算过程中，一直会分割出两个内部节点，这样的过程会被一再重复。因为每个生成的内部节点会继续分割出两个内部节点，其算法以所有叶节点（类别）错误率的加权总数作为剪枝的依据。

R 语言中可调用 rpart 包中的 rpart() 函数来运行分类与回归树。程序范例 6-1 中调用 rpart() 函数来分类鸢尾花（Iris）数据，鸢尾花数据中共有 5 个属性，前 4 个属性为自变量，分别为花萼长度（Sepal.Length）、花萼宽度（Sepal.Width）、花瓣长度（Petal.Length）及花瓣宽度（Petal.Width）。第 5 个属性 Species 为因变量（目标属性），其代表鸢尾花的 3 个品种：Setosa、Virginica 及 Versicolor。

[程序范例 6-1]

首先引用 rpart 包和 iris 数据。

```
> library(rpart)
> data(iris)
```

调用 sample() 函数随机抽取 10% 的观察值（数据笔数）作为测试数据：

```
> np = ceiling(0.1*nrow(iris))   # nrow(iris)返回iris数据的笔数
> np                              # ceiling()返回向上舍入的整数
[1] 15
> test.index = sample(1:nrow(iris),np)   # 10%作为测试数据
> iris.testdata = iris[test.index,]      # 测试数据
> iris.traindata = iris[-test.index,]    # 训练数据
```

调用 rpart()函数建立训练数据的决策树 iris.tree：

```
# 输出变量为Species
> iris.tree = rpart(Species ~ Sepal.Length + Sepal.Width +Petal.Length + Petal.Width, method="class", data=iris.traindata )
> iris.tree
n= 135
```

显示决策树 iris.tree 规则的信息（注意：若范例中使用随机抽取的数据，则它的运行结果可能不相同）：

```
>iris.tree

node), split, n, loss, yval, (yprob)
      * denotes terminal node

# 4条规则
1) root 135 88 virginica (0.3259259 0.3259259 0.3481481)
  2) Petal.Length< 2.45 44  0 setosa (1.0000000 0.0000000 0.0000000) *
  3) Petal.Length>=2.45 91 44 virginica (0.0000000 0.4835165 0.5164835)
    6) Petal.Width< 1.75 49  5 versicolor (0.0000000 0.8979592 0.1020408) *
    7) Petal.Width>=1.75 42  0 virginica (0.0000000 0.0000000 1.0000000) *
```

显示决策树 iris.tree 的 cp 值、错误率及各节点的详细信息：

```
> summary(iris.tree)
Call:
rpart(formula = Species ~ Sepal.Length + Sepal.Width + Petal.Length +
    Petal.Width, data = iris.traindata, method = "class")
  n= 135

          CP nsplit rel error    xerror       xstd
1 0.5056180      0 1.00000000 1.1910112 0.05361608
2 0.4382022      1 0.49438202 0.6629213 0.06475539
3 0.0100000      2 0.05617978 0.1235955 0.03571497

Variable importance
 Petal.Width Petal.Length Sepal.Length  Sepal.Width
          34           31           21           14

Node number 1: 135 observations,    complexity param=0.505618
  predicted class=setosa    expected loss=0.6592593  P(node) =1
    class counts:     46    45    44
   probabilities: 0.341 0.333 0.326
  left son=2 (46 obs) right son=3 (89 obs)
  Primary splits:
```

```
        Petal.Length < 2.45 to the left,  improve=45.49080, (0 missing)
        Petal.Width  < 0.8  to the left,  improve=45.49080, (0 missing)
        Sepal.Length < 5.45 to the left,  improve=31.08528, (0 missing)
        Sepal.Width  < 3.05 to the right, improve=17.89879, (0 missing)
    Surrogate splits:
        Petal.Width  < 0.8  to the left,  agree=1.000, adj=1.000, (0 split)
        Sepal.Length < 5.45 to the left,  agree=0.919, adj=0.761, (0 split)
        Sepal.Width  < 3.35 to the right, agree=0.830, adj=0.500, (0 split)

Node number 2: 46 observations
  predicted class=setosa     expected loss=0  P(node) =0.3407407
    class counts:     46     0     0
   probabilities: 1.000 0.000 0.000

Node number 3: 89 observations,    complexity param=0.4382022
  predicted class=versicolor expected loss=0.494382 P(node) =0.6592593
    class counts:      0    45    44
   probabilities: 0.000 0.506 0.494
  left son=6 (48 obs) right son=7 (41 obs)
  Primary splits:
      Petal.Width  < 1.75 to the left,  improve=35.209830, (0 missing)
      Petal.Length < 4.75 to the left,  improve=33.934380, (0 missing)
      Sepal.Length < 6.15 to the left,  improve=10.447780, (0 missing)
      Sepal.Width  < 2.95 to the left,  improve= 3.518872, (0 missing)
  Surrogate splits:
      Petal.Length < 4.75 to the left,  agree=0.899, adj=0.780, (0 split)
      Sepal.Length < 6.15 to the left,  agree=0.730, adj=0.415, (0 split)
      Sepal.Width  < 2.95 to the left,  agree=0.674, adj=0.293, (0 split)

Node number 6: 48 observations
  predicted class=versicolor expected loss=0.08333333 P(node) =0.3555556
    class counts:     0    44     4
   probabilities: 0.000 0.917 0.083

Node number 7: 41 observations
  predicted class=virginica  expected loss=0.02439024 P(node) =0.3037037
    class counts:     0     1    40
   probabilities: 0.000 0.024 0.976
```

画出决策树并标示文字：

```
> plot(iris.tree) ; text(iris.tree)
```

结果如图 6-2 所示。

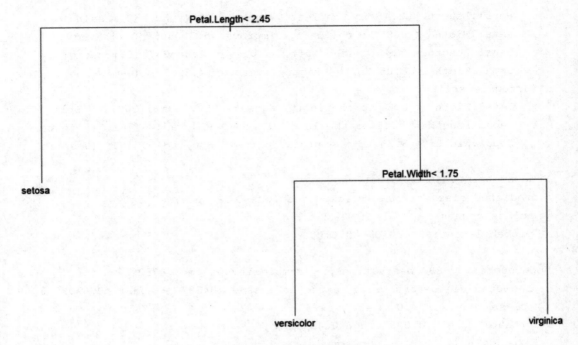

图 6-2 iris 的树结构图

显示训练数据的准确率:

```
# species.traindata是训练数据的真实输出变量
> species.traindata = iris$Species[-test.index]
# train.predict是训练数据的输出预测值
> train.predict=factor(predict(iris.tree, iris.traindata,
+ type='class'), levels=levels(species.traindata))

# 生成训练数据的混淆矩阵
> table.traindata =table(species.traindata,train.predict)
> table.traindata
                 train.predict
species.traindata setosa versicolor virginica
       setosa         47          0         0
       versicolor      0         42         1
       virginica       0          5        40
```

由以上表格可知:47 笔 setosa 都分类正确,有 1 笔 versicolor 错分至 virginica,有 5 笔 virginica 错分至 versicolor,所以准确率为(47+42+40)/ 135×100% = 95.55556%。

```
> correct.traindata=sum(diag(table.traindata))/sum(table.traindata)*100
                 # diag()返回矩阵对角线元素的值,sum()返回加总值
> correct.traindata
[1] 95.55556         # 训练数据的准确率为95.55556%
```

显示测试数据的准确率：

```
# species.testdata是测试数据的真实输出变量
> species.testdata = iris$Species[test.index]
# test.predict是测试数据的输出预测值
> test.predict=factor(predict(iris.tree, iris.testdata,
+ type='class'), levels=levels(species.testdata))

# 生成测试数据的混淆矩阵
> table.testdata = table(species.testdata,test.predict)
> table.testdata
                test.predict
species.testdata setosa versicolor virginica
      setosa         7         0         0
      versicolor     0         3         0
      virginica      0         1         4
>
```

由以上表格可知：测试数据共有 15 笔数据（观察值），setosa 和 versicolor 都分类正确，有 1 笔 virginica 错分至 versicolor，所以准确率为 (7+3+4) / 15×100% = 93.33333%。

```
> correct.testdata=sum(diag(table.testdata))/sum(table.testdata)*100
# diag()返回矩阵对角线元素的值，sum()返回加总值
> correct.testdata
[1] 93.33333        # 测试数据的准确率为93.33333%
```

[程序范例 6-2]

可设置 rpart.control()函数的自变量来改善决策树的分类结果，例如 rpart.control（minsplit=5, cp=0.0001, maxdepth=30）。

```
> library(rpart)
> data(iris)
>
> np = ceiling(0.1*nrow(iris))           # 10%作为测试数据
> np
[1] 15
>
> test.index = sample(1:nrow(iris),np)
>
> iris.testdata = iris[test.index,]      # 测试数据
> iris.traindata = iris[-test.index,]    # 训练数据
>
> iris.tree = rpart(Species ~ Sepal.Length + Sepal.Width +Petal.Length +
Petal.Width, method="class", data=iris.traindata,
```

```
+ control=rpart.control(minsplit=5, cp=0.0001, maxdepth=30) )
>
> species.traindata = iris$Species[-test.index]
> train.predict=factor(predict(iris.tree, iris.traindata,
+ type='class'), levels=levels(species.traindata))

# 生成训练数据的混淆矩阵
> table.traindata =table(species.traindata,train.predict)
> table.traindata
                  train.predict
species.traindata setosa versicolor virginica
       setosa         47          0         0
       versicolor      0         44         1
       virginica       0          2        41
> correct.traindata=sum(diag(table.traindata))/sum(table.traindata)*100
> correct.traindata
[1] 97.77778                # 训练数据的准确率为97.77778%
>
> species.testdata = iris$Species[test.index]
> test.predict=factor(predict(iris.tree, iris.testdata,
+ type='class'), levels=levels(species.testdata))
# 生成测试数据的混淆矩阵
> table.testdata =table(species.testdata,test.predict)
> table.testdata
                 test.predict
species.testdata setosa versicolor virginica
       setosa         3          0         0
       versicolor     0          5         0
       virginica      0          0         7
> correct.testdata=sum(diag(table.testdata))/sum(table.testdata)*100
> correct.testdata
[1] 100                     # 测试数据的准确率为100%
>
> iris.tree
n= 135

node), split, n, loss, yval, (yprob)
      * denotes terminal node

# 产出5条规则
 1) root 135 88 setosa (0.34814815 0.33333333 0.31851852)
   2) Petal.Length< 2.45 47  0 setosa (1.00000000 0.00000000 0.00000000) *
   3) Petal.Length>=2.45 88 43 versicolor (0.00000000 0.51136364 0.48863636)
     6) Petal.Width< 1.75 49  5 versicolor (0.00000000 0.89795918 0.10204082)
```

```
 12) Petal.Length< 4.95 43   1 versicolor (0.00000000 0.97674419 0.02325581) *
 13) Petal.Length>=4.95 6    2 virginica  (0.00000000 0.33333333 0.66666667)
   26) Petal.Width>=1.55 3   1 versicolor (0.00000000 0.66666667 0.33333333) *
   27) Petal.Width< 1.55 3   0 virginica  (0.00000000 0.00000000 1.00000000) *
  7) Petal.Width>=1.75 39    1 virginica  (0.00000000 0.02564103 0.97435897) *
```

ID3 使用信息增益（Information Gain）作为属性选择指标（Selection Measure），而 C5.0 是 Quinlan 学者依据 ID3 算法改进而来的。C5.0 和 ID3 一样是以信息增益最大的属性作为分割属性，两者最大的差异在于 ID3 偏向选择属性值较多的属性，而 C5.0 使用增益比例（Gain Ratio）作为属性选择指标。信息增益的计算方式如下：

$$\text{Info}(S) = -\sum_{i=1}^{k} \left\{ \left[\text{freq}(C_i, S/|S|) \right] \log_2 \left[\text{freq}(C_i, S/|S|) \right] \right\} \quad (6\text{-}1)$$

$$\text{Info}_x(S) = -\sum_{i=1}^{L} \left[(|S_i|/|S|) \text{Info}(S_i) \right] \quad (6\text{-}2)$$

$$\text{Gain}(X) = \text{Info}(S) - \text{Info}_x(S) \quad (6\text{-}3)$$

$$\text{SplitInfo}(X) = -\sum_{i=1}^{L} \left[\frac{|S_i|}{|S|} \text{Log}_2 \frac{|S_i|}{|S|} \right] \quad (6\text{-}4)$$

$$\text{GainRatio}(X) = \text{Gain}(X)/\text{SplitInfo}(X) \quad (6\text{-}5)$$

在公式（6-1）中，$|S|$ 是在训练数据中各类别的总个数；c 为目标属性中所包含的类别，S_i 为类别 i 下包含的笔数；在公式（6-2）中，$|S|$ 为选定属性下训练数据中各类别的总个数，L 为输入的属性数量，$\text{Info}_x(S)$ 代表在某一情况下选择一个候选属性的熵值（Entropy）；在公式（6-3）中，$\text{Gain}(X)$ 代表信息增益，$\text{Info}(S)$ 为未增加某一属性时的熵值。C5.0 初始时利用公式（6-1）计算不考虑属性时的信息复杂度；公式（6-2）为考虑某一属性情况时的信息复杂度；公式（6-3）为公式（6-1）与公式（6-2）相减后的信息复杂度的差值，即为信息增益；公式（6-4）为在某一情况下分割节点的熵值；公式（6-5）为其增益比例。

在建立决策树的过程中，决策树算法利用最小案例数量（Minimum Case，M）检验每个节点所包含的案例数量是否超过 M，若没超过，则该节点停止生长；若超过，则继续往下生长。当树结构生长完后，为了避免造成过拟合问题（即降低树的复杂度），需对树进行剪枝。C5.0 采用后剪枝法，在剪枝阶段利用设置剪枝的信赖水平（Confidence Level，CF）下的信赖区间作为基准，计算该内部节点以及下一节点在统计上的预期错误率与案例数的信赖水平，当目前内部节点错误率小于下一个内部节点的错误率时，若统计显示该节点继续分割下去也不会有更好的结果，便将其剪掉。

[程序范例6-3]

首先引用 C50 包和 iris 数据。

```
> library(C50)
> data(iris)
```

```
> np = ceiling(0.1*nrow(iris))           # 10% 作为测试数据
> np
[1] 15
```

随机取用10%作为测试数据、90%作为训练数据：

```
# 抽样
> test.index = sample(1:nrow(iris),np)
> iris.test = iris[test.index,]          # 测试数据
> iris.train = iris[-test.index,]        # 训练数据
```

设置C5.0相关自变量：

```
# sample=0表示不重新抽样
> c=C5.0Control(subset = FALSE,
+               bands = 0,
+               winnow = FALSE,
+               noGlobalPruning = FALSE,
+               CF = 0.25,
+               minCases = 2,
+               fuzzyThreshold = FALSE,
+               sample = 0,
+               seed = sample.int(4096, size = 1) - 1L,
+               earlyStopping = TRUE
+               )
# 第5个属性为目标属性(输出变量)
> iris_treeModel <- C5.0(x = iris.train[, -5], y = iris.train$Species,control
=c)
```

显示C5.0决策树iris.treeModel的规则及训练数据的错误率：

```
> summary(iris_treeModel)
Call:
C5.0.default(x = iris.train[, -5], y = iris.train$Species, control = c)


C5.0 [Release 2.07 GPL Edition]
-------------------------------

Class specified by attribute `outcome'

Read 135 cases (5 attributes) from undefined.data

Decision tree:

Petal.Length <= 1.7: setosa (41)
Petal.Length > 1.7:
```

```
:...Petal.Width > 1.7: virginica (43/1)
   Petal.Width <= 1.7:
   :...Petal.Length <= 5.3: versicolor (49/1)
      Petal.Length > 5.3: virginica (2)


Evaluation on training data (135 cases):

         Decision Tree
       ----------------
       Size      Errors

          4     2( 1.5%)   <<


        (a)   (b)   (c)    <-classified as
       ----  ----  ----
         41                (a): class setosa
               48     1    (b): class versicolor
                1    44    (c): class virginica


    Attribute usage:

    100.00% Petal.Length
     69.63% Petal.Width


Time: 0.0 secs
```

由以上结果可知，训练数据的错误率为 1.5%（准确率为 98.5%）。

```
# 生成测试数据的输出
> test.output=predict(iris_treeModel, iris.test[, -5], type = "class")
> n=length(test.output)
> number=0
> for( i in 1:n)                # 计算测试数据的准确率
+ {
+   if (test.output[i] == iris.test[i,5])
+   {
+     number=number+1
+   }
+ }
> test.accuracy=number/n*100
> test.accuracy
```

```
[1] 73.33333
```

由以上结果可知，测试数据的准确率为73.33333%。

需要特别注意的是，C5.0的输出变量（例如 iris.train$Species）的数据类型必须是因子，若非因子数据类型，则可调用 factor()函数来转换。

```
> iris.train$Species=factor(iris.train$Species)
```

[程序范例6-4]

我们也可以直接设置 C5.0Control()函数的 sample 自变量来代表训练数据笔数，例如 sample=0.9 表示 90%作为训练数据。

```
> library(C50)
> library(stringr)
>
> data(iris)

# 使用 sample = 0.9 表示90%作为训练数据
> c=C5.0Control(subset = FALSE,
+               bands = 0,
+               winnow = FALSE,
+               noGlobalPruning = FALSE,
+               CF = 0.25,
+               minCases = 2,
+               fuzzyThreshold = FALSE,
+               sample = 0.9,              # 90%作为训练数据
+               seed = sample.int(4096, size = 1) - 1L,
+               earlyStopping = TRUE,
+               label = "Species")
> iris_treeModel <- C5.0(x = iris[, -5], y = iris$Species,
+              control =c)
>
> summary(iris_treeModel)

Call:
C5.0.default(x = iris[, -5], y = iris$Species, control = c)


C5.0 [Release 2.07 GPL Edition]
-------------------------------

Class specified by attribute `Species'

Read 135 cases (5 attributes) from undefined.data
```

```
Decision tree:

Petal.Length <= 1.9: setosa (45)
Petal.Length > 1.9:
:...Petal.Width > 1.7: virginica (41/1)
    Petal.Width <= 1.7:
    :...Petal.Length <= 4.9: versicolor (43/1)
        Petal.Length > 4.9: virginica (6/2)
Evaluation on training data (135 cases):

        Decision Tree
        ----------------
        Size      Errors

          4      4( 3.0%)   <<


         (a)   (b)   (c)    <-classified as
        ----  ----  ----
          45                (a): class setosa
                42     3    (b): class versicolor
                 1    44    (c): class virginica



    Attribute usage:

    100.00% Petal.Length
     66.67% Petal.Width


Evaluation on test data (15 cases):

        Decision Tree
        ----------------
        Size      Errors

          4      0( 0.0%)   <<


         (a)   (b)   (c)    <-classified as
        ----  ----  ----
           5                (a): class setosa
                 5          (b): class versicolor
```

```
               5     (c): class virginica

Time: 0.0 secs
```

调用stringr包中的str_locate_all()函数和substr()函数得到测试数据的错误率。首先使str_locate_all()函数取得列表iris_treeModel的output元素（iris_treeModel$output）中<<的位置，再由substr()函数取得%前1位置至前4位置的文字（表示为测试数据错误率），最后调用as.numeric()函数转换为数字。

```
> tt=as.character(iris_treeModel$output)   # 转换文字
> x=str_locate_all(tt,"<<")
> y=substr(tt,x[[1]][2]-9,x[[1]][2]-6)

> test.error=as.numeric(y)
> test.correct=100-test.error
> test.correct              # 测试数据准确率
[1] 100
```

6.2 支持向量机

支持向量机（Support Vector Machine，SVM）是由Vapnik在1995年根据AT&T实验室团队所提出的统计学习理论（Statistical Learning Theory）发展出来的学习算法。支持向量机的学习架构是以小样本的训练数据学习来得到最佳的学习与归纳能力，使得支持向量机能够学习出较为平滑的曲线，测试数据也能在最少的变化中进行分类（Classification）或回归（Regression），即结构风险最小化（Structure Risk Minimization，SRM）原则。

以图6-3的二维空间为例，其中白点与黑点分别代表两类训练数据。对于H1、H2、H3三条线，每一条线都可视为一个分类器，我们可以找出H2分类器、使这两类产生最大的距离，借以得到最小的分类错误率，故H2为此图中最佳的分类器。在高维度空间中，支持向量机可找出一个超平面（Hyper-Plane）分类器，借以得到最小的分类错误率。支持向量机在解决非线性分类问题方面也有极佳的处理能力。

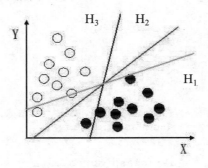

图6-3　分类示意图

对于分类的问题，其实就是寻找一个函数 $y = f(x)$，$x \in R_n$，$y \in \{1, -1\}$，在给定的 k 个训练数据 $(x_1, y_1) \sim (x_i, y_i)$，$i = 1 \sim k$ 中，再寻找一个最佳超平面，可将训练数据加以分类，计算公式如下：

$$(w \cdot x) + b = 0, w \in R^n, b \in R \tag{6-6}$$

此平面满足下列条件：

$$y_i[w \cdot x_i + b] > 0, i = 1 \sim k \tag{6-7}$$

此时，为了避免因为噪声而使得误差过大，必须在此加入松弛变量 $\zeta_i \geqslant 0$，使得计算公式变成：

$$y_i[w \cdot x_i + b] > 1 - \zeta_i, i = 1 \sim k \tag{6-8}$$

此时被超平面分割成两部分的训练数据，其各自到达超平面的最小距离为最大，计算方法如下：

$$\text{Min}: [\gamma(w) = \frac{1}{2} w \cdot w + C \sum_{i=1}^{k} \zeta_i] \tag{6-9}$$

此有限制的优化问题可以通过 Lagrange 乘数（Multiplier）转化为对偶问题：

$$L = \frac{1}{2} \|w\|^2 + C \sum_{i=1}^{k} \zeta_i - P - \sum_{i=1}^{k} \gamma_i \zeta_i \tag{6-10}$$

$$P = \sum_{i=1}^{k} \alpha \{y_i(w \cdot x_i + b) - 1 + \zeta_i\} \tag{6-11}$$

其中，L 包含 4 个参数 (w, b, α, γ)，需满足对 w、b 最小化且对 α、γ 最大化，而经由 Karush-Kuhn-Tucker（KKT）理论即可求得：$\sum_{i=1}^{k} \alpha_i y_i = 0$，$w = \sum_{i=1}^{k} \alpha_i y_i x_i$，由此可得到最大化函数：

$$\text{Max}: \left[\sum_{i=1}^{k} \alpha_i - \frac{1}{2} \sum_{i,j=1}^{k} \alpha_i \alpha_j y_i y_j (x_i \cdot x_j) \right] \tag{6-12}$$

限制条件为：

$$0 \leqslant a_i \leqslant C, i = 1 \sim k, \sum_{i=1}^{k} a_i y_i = 0$$

支持向量机可将数据由 $k(u, v) = (\Phi(x_i), \Phi(x_j))$ 转换至高维度的特征空间中，所以支持向量机处理优化的问题可以转变为：

$$L_D = \sum_{i=1}^{m} \alpha_i - \frac{1}{2} \sum_{i,j=1}^{m} \alpha_i \alpha_j y_i y_j (\Phi(x_i), \Phi(x_j)) \tag{6-13}$$

$$L_D = \sum_{i} \alpha_i - \frac{1}{2} \sum_{i,j} \alpha_i \alpha_j y_i y_j k(u, v) \tag{6-14}$$

限制条件为：

$$0 \leq a_i \leq C, i = 1 \sim k, \sum_{i=1}^{k} a_i y_i = 0$$

其中 $k(u, v)$ 称为核函数，常用的核函数包含 linear、polynomial、radial base 及 sigmoid 函数，linear 函数定义为 $k(u, v) = u'v$，polynomial 函数定义为 $k(u, v) = (u'v + coef0)^{\deg ree}$，radial base 函数定义为 $k(u, v) = e^{(-\gamma|u-v|^2)}$，sigmoid 函数定义为 $k(u, v) = \tanh(ru'v + coef0)$。

[程序范例6-5]

首先引用 e1071 包和 iris 数据：

```
> library(e1071)
> data(iris)
> index <- 1:nrow(iris)
> np = ceiling(0.1*nrow(iris))       # 10%作为测试数据
> np
[1] 15
```

随机取用 10%作为测试数据、90%作为训练数据：

```
> testindex = sample(1:nrow(iris),np)
> testset = iris[test.index,]        # 测试数据
> trainset = iris[-test.index,]      # 训练数据
```

使用训练数据建立 svm.model 模型，核函数为 radial base 并设置自变量 cost（C）=10、gamma（γ）=10：

```
# 输出变量是Species, .表示使用除输出变量外的变量作为输入变量
> svm.model <- svm(Species ~ ., data = trainset, type = 'C-classification', cost = 10, gamma = 10)
```

使用 svm.model 模型来预测测试数据：

```
> svm.pred <- predict(svm.model, testset[,-5])
```

显示测试数据的准确率：

```
# 生成测试数据的混淆矩阵
> table.svm.test=table(pred = svm.pred, true = testset[,5])
> table.svm.test
            true
pred         setosa versicolor virginica
  setosa          2          0         0
  versicolor      0          7         0
  virginica       1          1         4

# 对角线上的值表示正确数量
> correct.svm=sum(diag(table.svm.test))/sum(table.svm.test)
```

```
> correct.svm=correct.svm*100
> correct.svm
[1] 86.66667                           # 测试数据的准确率
```

e1071 包提供了 tune.svm()函数搜索最佳的 cost(C)和 gamma(γ)值。我们可设置搜索范围，例如 $0.001 <= \gamma <= 0.1$ 及 $0.1 <= C <= 10$。

```
# 0.001<=γ<=0.1; 0.1<=C<=10
> tuned <- tune.svm(Species ~., data = trainset, gamma = 10^(-3:-1), cost = 10^(-1:1))
> summary(tuned)

Parameter tuning of 'svm':

- sampling method: 10-fold cross validation

- best parameters:
 gamma cost
   0.1   10

- best performance: 0.04450549

- Detailed performance results:
  gamma cost      error dispersion
1 0.001  0.1 0.74395604 0.13837721
2 0.010  0.1 0.42362637 0.22152013
3 0.100  0.1 0.14065934 0.08211993
4 0.001  1.0 0.41648352 0.22753121
5 0.010  1.0 0.13351648 0.07800779
6 0.100  1.0 0.05879121 0.05843897
7 0.001 10.0 0.13351648 0.07800779
8 0.010 10.0 0.05219780 0.06125272
9 0.100 10.0 0.04450549 0.06261755
```

由以上结果可知，$\gamma = 0.1$ 和 C=10 误差值最小，所以选定此值作为 svm()函数的自变量并重新计算准确率。

```
# γ=0.1; C=10
> model  <- svm(Species ~., data = trainset, kernel="radial", gamma=0.1, cost=10)
> summary(model)

Call:
svm(formula = Species ~ ., data = trainset, kernel = "radial",
    gamma = 0.1, cost = 10)


Parameters:
```

```
     SVM-Type:  C-classification
   SVM-Kernel:  radial
         cost:  10
        gamma:  0.1

Number of Support Vectors:  31

 ( 3 15 13 )
Number of Classes:  3

Levels:
 setosa versicolor virginica

# 生成测试数据的预测值
> svm.pred <- predict(model, testset[,-5])
>
# 生成测试数据的混淆矩阵
> table.svm.best.test=table(pred = svm.pred, true = testset[,5])
> table.svm.best.test
            true
pred         setosa versicolor virginica
  setosa          3          0         0
  versicolor      0          8         0
  virginica       0          0         4
>
correct.svm.best=sum(diag(table.svm.best.test))/sum(table.svm.best.test)*100
> correct.svm.best
[1] 100                     # 测试数据的准确率为100%
```

6.3 人工神经网络

人工神经网络是由许多人工神经细胞相互连接所组成的,并且可以组成各种网络模型（Network Model）。人工神经细胞又称为人工神经元、处理单元。处理单元（Processing Element, PE）是人工神经网络中最基本的组成单位,每一个处理单元的输出必须连接到下一层的处理单元。处理单元的输入、输出值计算公式如下：

$$f_i = f(\text{net}_i) = f\left(\sum_i W_{ij} X_i - \theta_i\right) \tag{6-15}$$

其中：

- f_i：人工神经网络处理单元的输出信号。
- net_i：集成函数。

- f：人工神经网络处理单元的转换函数。
- W_{ij}：人工神经网络各处理单元间的连续权重值（Weight）。
- X_i：输入向量。
- θ_i：人工神经网络处理单元的阈值（Threshold）。

反向传播（Back Propagation，BP）人工神经网络是被广泛应用的一种监督式学习的人工神经网络，它的基本原理是利用最陡坡降法（Gradient Steepest Descent Method）以迭代方式将误差函数最小化。反向传播算法的网络训练包含两个阶段：前向传播阶段和反向传播阶段，如图 6-4 所示。前向传播（也称为正向传播）阶段时输入向量从输入层开始输入，并以前向方式经由隐藏层传至输出层，最后推导出输出值。在此阶段，网络节点间的权重值都是固定的。反向传播阶段，网络节点间的连接权重值则根据误差修正规则来进行修正，并借助连接权重值的修正，使修正后的推导输出值偏向于目标输出值。误差函数公式如下：

$$E = \frac{1}{2}\sum_j (d_j - y_j)^2 \quad (6\text{-}16)$$

其中：
- d_j：输出层第 j 个输出单元的目标输出值。
- y_j：输出层第 j 个输出单元的推导输出值。

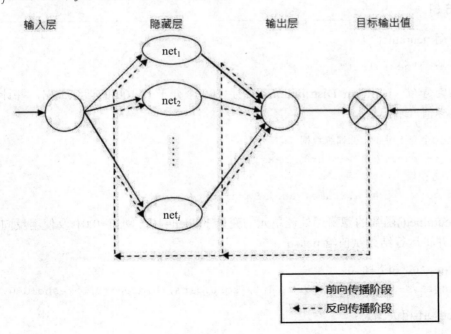

图 6-4　反向传播算法的前向传播阶段和反向传播阶段

误差函数对权重值的偏微分公式表示如下：

$$\Delta W = -\eta \frac{\partial E}{\partial W} \quad (6\text{-}17)$$

其中：

- ΔW：代表各层处理单元间的连接权重值的修正量。
- η：学习率（Learning Rate），主要用来控制每次权重值修改的大小。

误差函数对隐藏层第 k 个单元与输出层第 j 个单元间的连接权重值关系，公式如下：

$$\Delta W_{kj} = \eta \times \delta_j \times X_k \tag{6-18}$$

其中：

- X_k：第 k 个隐藏层单元的输入向量。
- δ_j：区域梯度函数，$\delta_j = (d_j - y_j) y_j (1 - y_j)$。

误差函数对输入层第 i 个单元与隐藏层第 k 个单元间的连接权重值关系，公式如下：

$$\Delta W_{ik} = \eta \times \delta_k \times X_i \tag{6-19}$$

其中：

- δ_k：区域梯度函数，$\delta_k = y_k (1 - y_k) \sum_i \delta_i W_{ik}$。

[程序范例6-6]

首先引用 neuralnet 包：

```
> library("neuralnet")
```

使用均匀分布（Uniform Distribution）产生 100 个介于 0~100 的训练数据，并计算其平方根值作为目标值（输出值）：

```
# 产生100个介于0~100的训练数据
> Var1 <- runif(100, min=0, max=100)
# 计算平方根值
> sqrt.data <- data.frame(Var1, Sqrt=sqrt(Var1))
```

设置 neuralnet() 函数的隐藏层处理单元的变量 hidden=10、阈值=0.01 及使用反向传播人工神经网络，并将运行结果赋值给 net.sqrt：

```
# hidden=10，阈值=0.01
> net.sqrt <- neuralnet(Sqrt~Var1, sqrt.data, hidden=10,threshold=0.01)
```

打印 net.sqrt 相关信息：

```
> print(net.sqrt)
Call: neuralnet(formula = Sqrt ~ Var1, data = sqrt.data, hidden = 10,     threshold = 0.01)

1 repetition was calculated.

          Error Reached Threshold Steps
```

1 0.0001658715951   0.009936046689 10537

画出 net.sqrt 的架构图，如图 6-5 所示。

```
> plot(net.sqrt)
```

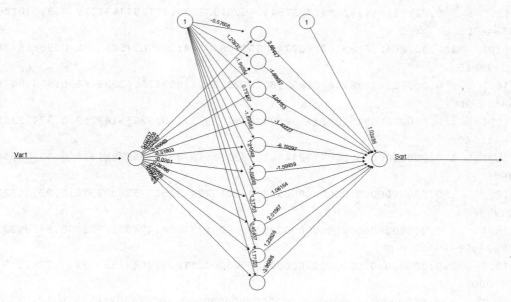

图 6-5　net.sqrt 的架构图

产生测试数据并调用 compute() 函数来预测此模型的结果：

```
# 产生测试数据
> testdata <- as.data.frame((1:10)^2)
> nn.result <- compute(net.sqrt, testdata)
```

打印预测结果：

```
> print(nn.result)
$neurons
$neurons[[1]]
      1  (1:10)^2
 [1,] 1         1
 [2,] 1         4
 [3,] 1         9
 [4,] 1        16
 [5,] 1        25
 [6,] 1        36
 [7,] 1        49
 [8,] 1        64
 [9,] 1        81
[10,] 1       100
```

```
$neurons[[2]]
                                                                      [,2]          [,3]              [,4]
     [1,]    1 0.25772581220548373970302691304823383688926696977734 0.9919478656
0.6044046623
     [2,]    1 0.02292984448268997573430960613222850952297449111  0.9907018379
0.8125646725
     [3,]    1 0.00026312618946315027498861205401681218063458800 32 0.9881873662
0.9610199115
     [4,]    1 0.00000048978148008862760266236835438036223422386 68 0.9835047641
0.9964540598
     [5,]    1 0.00000000015123349194929222940892571402926591872 53 0.9747324334
0.9998442540
     [6,]    1 0.00000000000000774845630382309519748607051781164 0  0.9577294311
0.9999965980
     [7,]    1 0.00000000000000000658722354102799925127526003976    0.9235472752
0.9999999629
     [8,]    1 0.00000000000000000000009292047371886474438178 18   0.8539429308
0.9999999998
     [9,]    1 0.0000000000000000000000000000021749117047258793 9  0.7197913330
1.0000000000
     [10,]   1 0.00000000000000000000000000000000000008446816267  0.5060569487
1.0000000000
                              [,5]              [,6]          [,7]                [,8]
     [1,]  0.802116690204665583 0.7968909277964 0.2510113653
0.55728999915289978428490
     [2,]  0.675954358580697079 0.7449406210565 0.2677665348
0.32812303045151075542307
     [3,]  0.408061586830988920 0.6410360878618 0.2972179318
0.09155600904400451600651
     [4,]  0.127627524893795391 0.4728123320481 0.3414099410
0.01094230349795086154607
     [5,]  0.019548653245394374 0.2700522148165 0.4024439148
0.00064559742156132342555
     [6,]  0.001741956148366336 0.1113886025399 0.4811573388
0.00002006528651787095071
     [7,]  0.000098064227900182 0.0337093301416 0.5750879525
0.00000033152970991952383
     [8,]  0.000003539693612293 0.0079112815363 0.6767441203
0.00000000291371996069367
     [9,]  0.000000082042026592 0.0014949996732 0.7743813690
0.00000000001362165003402
     [10,] 0.000000001221126965 0.0002308477486 0.8564052781
0.00000000000003387414335
                       [,9]           [,10]          [,11]
```

```
 [1,]  0.541182501700  0.313699456493  0.3014415717
 [2,]  0.504031573736  0.288055996536  0.3211166332
 [3,]  0.442217067074  0.248221939978  0.3553398072
 [4,]  0.358982541606  0.198976939161  0.4057762028
 [5,]  0.263721441826  0.146966628131  0.4735095293
 [6,]  0.171794037122  0.099231314426  0.5573786856
 [7,]  0.098098407455  0.060979131897  0.6521084818
 [8,]  0.049106005126  0.034088025562  0.7478745784
 [9,]  0.021718638784  0.017373953135  0.8330633738
[10,]  0.008567711444  0.008100501284  0.8992465457

$net.result
            [,1]
 [1,]  1.029557458
 [2,]  2.000101230
 [3,]  2.998497421
 [4,]  4.001433416
 [5,]  4.998696195
 [6,]  6.001244074
 [7,]  6.999690285
 [8,]  7.998276198
 [9,]  9.002924905
[10,]  9.990079622
```

若要计算平均绝对误差（mae）及均方根误差（rmse），则可再加载 DMwR 包并调用 regr.eval() 函数：

```
> library(DMwR)

# 计算平均绝对误差(mae)及均方根误差(rmse)
> regr.eval(expected.output,nn.result$net.result[,1],
+ stats=c('mae','rmse'))

        mae           rmse
0.05002136169  0.03148412327
```

6.4 集成学习方法

  Nilsson 在 1965 年提出的方法，由多位专家组合而成，按一些特定方式（如投票法、权重法）整合各专家的意见进行决策，得到的结果比只采纳单个专家意见进行决策的效果更好；由于每位专家擅长的领域不同，通过集成学习的机制可以让专家彼此互补，因而可以得到更好的结果。常用的集成学习方法包含装袋法（Bagging）及提升法（Boosting）。

6.4.1 随机森林

装袋法是将所有预测模型的多数预测值作为未知值组的预测值，主要采用放回（With Replacement）随机抽样方式，从原始数据集合中选取与原集合相同数量的数据，产生多个训练集合，使用投票法并按得票多来决定分类结果。

随机森林（Random Forest）是一个特别设计给决策树分类法使用的装袋法，它结合多个决策树的预测结果，每棵树都是根据随机森林的随机向量的值来建立的。

[程序范例 6-7]

首先清除 R 软件内存中的数据（确认没有上次使用过的数据对象）：

```r
# 清除对象(变量)
> rm(list = ls())
> gc()
```

引用 randomForest 包和 iris 数据集：

```r
> library(randomForest)
> data(iris)
```

调用 sample() 函数将 iris 数据分为训练数据集（80%）和测试数据集（20%）：

```r
# 抽样
> ind <- sample(2, nrow(iris), replace=TRUE, prob=c(0.8, 0.2))
> trainData <- iris[ind==1,]
> testData <- iris[ind==2,]
```

使用训练集数据和 100 棵决策树建立随机森林：

```r
> rf <- randomForest(Species ~ ., data=trainData, ntree=100)
```

调用 predict() 函数来预测测试数据的准确率：

```r
# 产生测试数据预测值
> irisPred <- predict(rf, newdata=testData)
> table(irisPred, testData$Species)

irisPred     setosa versicolor virginica
  setosa         13          0         0
  versicolor      0          9         0
  virginica       0          0         9
```

由以上数据可知，测试数据的准确率为 100%。

6.4.2 提升法

提升法是对每个训练数据集设置一个权重，每次迭代后，对分类错误的数据加大权重，使得下一次的迭代更加关注这些数据。提升法采用不放回随机抽样。

[程序范例 6-8]

引用 adabag 包和 iris 数据集：

```
> library(adabag)
> data(iris)
```

调用 sample()函数将 iris 数据分为训练数据集（80%）及测试数据集（20%）：

```
# 抽样
> ind <- sample(2, nrow(iris), replace=FALSE, prob=c(0.8, 0.2))
> trainData <- iris[ind==1,]
> testData <- iris[ind==2,]
```

使用训练集数据并迭代 5 次来建立提升法：

```
# mfinal表示迭代次数
> train.adaboost <- boosting(Species~., data=trainData, boos=TRUE, mfinal=5)
```

调用 predict.boosting()函数来预测测试数据的准确率：

```
> test.adaboost.pred <- predict.boosting(train.adaboost,newdata=testData)

# 生成测试数据的混淆矩阵
> test.adaboost.pred$confusion
               Observed Class
Predicted Class setosa versicolor virginica
     setosa          7          0         0
     versicolor      0          9         0
     virginica       0          1         9

# 产生error值
> test.adaboost.pred$error
[1] 0.03846154
```

由以上数据可知，测试数据的准确率为 25/26×100=96.15%（100 − 3.85 = 96.15）。

6.5 习题

（1）使用决策树分析 bank.csv。
（2）使用其他分类方法分析 bank.csv。

第 7 章

非监督式学习

本章介绍常用的非监督式学习（聚类）算法及其应用，例如层次聚类法（Hierarchical Clustering）、K 均值聚类算法（K Means）及模糊 C 均值聚类算法（Fuzzy C Means）。

7.1 层次聚类法

聚类（Clustering）是将一组数据依据相似度计算公式将其凝聚或分裂成若干簇（聚类）。层次聚类法是通过如图 7-1 所示的层次架构方式，将数据层层反复地进行凝聚或分裂，以产生最后的树结构。常见的方式有两种：凝聚法（Agglomerative）是采用自下向上的处理方式，从树结构的底部开始，将数据或各聚类逐次合并，一开始将每个数据都视为一个独立的聚类，然后根据聚类间的相似度计算公式不断合并两个最相似的聚类，直到最后所有的聚类都合并成一个大的类为止；分拆法（Divisive，或称为分裂法）是采用自上向下的处理方式，从树结构的顶端开始将大的聚类逐次分拆，一开始时将所有数据视为一个大的聚类，根据相似度计算公式不断地将大聚类分拆成较小的聚类，直到聚类数量达到事先所设置的数目为止。

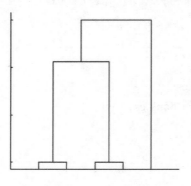

图 7-1 层次聚类法的树结构

层次聚类法的相似度可由距离来计算，两个聚类（c_i, c_j）相似度（Similarity）的计算公式如下：

$$\text{Similarity} = \frac{1}{1 + d(c_i, c_j)} \qquad (7\text{-}1)$$

其中两个聚类间的距离 $d(c_i, c_j)$ 可使用图 7-2 提供的 4 种方法来计算：单一连接法，即取聚类和聚类间最近点的距离 $d_{\min}(c_i, c_j)$；中心法，即取聚类和聚类间中心点的距离 $d_{\text{mean}}(c_i, c_j)$；平均法，即取聚类和聚类间所有点距离的平均 $d_{\text{avg}}(c_i, c_j)$；完全连接法，即取聚类和聚类间最远点的距离 $d_{\max}(c_i, c_j)$。对于点和点间的距离，则可使用：欧几里得距离（Euclidean Distance）、皮尔森相关系数（Pearson Correlation Coefficient）或马氏距离（Mahalanobis Distance）。

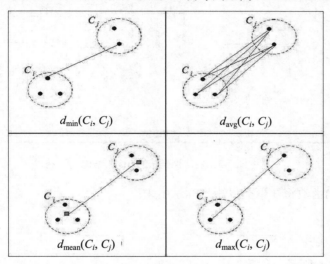

图 7-2 计算各聚类间的距离

[程序范例 7-1]

为了方便展示层次聚类法的树结构，本范例只随机取用 10%（15 笔）的 iris 数据作为样本：

```
> data(iris)
> index <- 1:nrow(iris)

# 使用10%的数据作为样本
> np <- ceiling(0.1*nrow(iris))
> idx <- sample(1:nrow(iris),np)
# 抽样
> irisSample <- iris[idx,]
```

由于层次聚类法是非监督式学习法，因此将 irisSample 目标属性 Species 设为 NULL：

```
# 不使用目标属性
> irisSample$Species <- NULL
```

调用 hclust() 函数并设置使用单一连接法来计算聚类距离：

```
# method="single"表示使用单一连接法
> hc <- hclust(dist(irisSample), method="single")
```

调用 plot() 函数绘出树结构并标示目标属性 Species 的品种，其结果如图 7-3 所示。

```
> plot(hc,labels=iris$Species[idx])
```

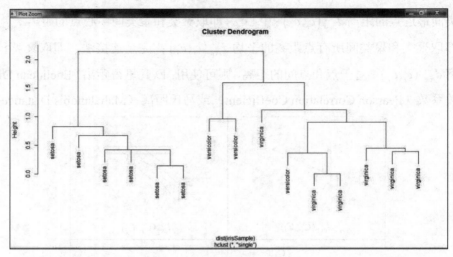

图 7-3　层次聚类法的 iris 树结构

调用 rect.hclust() 函数将 3 聚类以矩形标示出来：

```
# k=3
> rect.hclust(hc, k=3)
```

最后结果如图 7-4 所示。

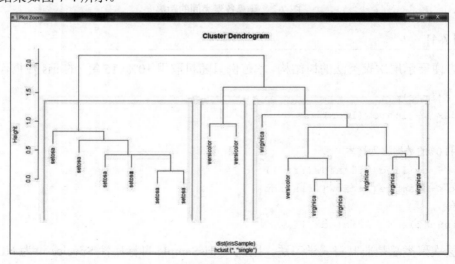

图 7-4　层次聚类法的 iris 树结构+矩形

7.2 K 均值聚类算法

K 均值聚类算法是 MacQueen 于 1967 年提出的聚类算法，必须事前设置聚类的数量为 k，k 个聚类 C_i，$i=1,2,\cdots,k$，计算公式（7-2），借以达到聚类优化的目的。

$$\mathrm{argmin}\sum_{i=1}^{k}\sum_{x_j \in C_i} x_j - \mu_i \tag{7-2}$$

其中：

μ_i：第 i 个聚类的聚类（簇）中心。

K 均值聚类算法依照下列 4 个步骤进行聚类：

（1）给定 k 值，将数据分割成 k 个非空子集合。
（2）现在划分聚类组的聚类簇心，这些聚类簇心为各个聚类组的中心点。
（3）将每个数据归类于最接近的聚类簇的中心点。
（4）回到步骤（2），一直到每个聚类组的数据没有任何变化。

[程序范例 7-2]

首先载入 iris 数据并显示其属性：

```
> data(iris)

# 显示5个属性
> attributes(iris)
$names
[1] "Sepal.Length" "Sepal.Width"  "Petal.Length" "Petal.Width"  "Species"

$row.names
    [1]   1   2   3   4   5   6   7   8   9  10  11  12  13  14  15  16  17  18  19
 20  21  22  23  24  25  26  27  28  29
   [30]  30  31  32  33  34  35  36  37  38  39  40  41  42  43  44  45  46  47  48
 49  50  51  52  53  54  55  56  57  58
   [59]  59  60  61  62  63  64  65  66  67  68  69  70  71  72  73  74  75  76  77
 78  79  80  81  82  83  84  85  86  87
   [88]  88  89  90  91  92  93  94  95  96  97  98  99 100 101 102 103 104 105 106
107 108 109 110 111 112 113 114 115 116
  [117] 117 118 119 120 121 122 123 124 125 126 127 128 129 130 131 132 133 134
135 136 137 138 139 140 141 142 143 144 145
  [146] 146 147 148 149 150

$class
[1] "data.frame"
```

将 iris 赋值给 iris2 并将其目标属性设为 NULL：

```
> iris2 <- iris
> iris2$Species <- NULL
```

调用 kmeans()函数将 iris2 数据分为 3 聚类，并直接显示结果（注意：使用小括号括住表达式）：

```
# 使用小括号直接显示结果
> (kmeans.result <- kmeans(iris2, 3))
K-means clustering with 3 clusters of sizes 50, 38, 62

Cluster means:
  Sepal.Length Sepal.Width Petal.Length Petal.Width
1     5.006000    3.428000     1.462000    0.246000
2     6.850000    3.073684     5.742105    2.071053
3     5.901613    2.748387     4.393548    1.433871

Clustering vector:
  [1] 1 1 1 1 1 1 1 1 1 1 1 1 1 1 1 1 1 1 1 1 1 1 1 1 1 1 1 1 1
 [30] 1 1 1 1 1 1 1 1 1 1 1 1 1 1 1 1 1 1 1 1 1 3 3 2 3 3 3 3 3
 [59] 3 3 3 3 3 3 3 3 3 3 3 3 3 3 3 2 3 3 3 3 3 3 3 3 3 3 3 3 3
 [88] 3 3 3 3 3 3 3 3 3 3 3 3 3 2 3 2 2 2 2 2 3 2 2 2 2 2 2 3 3 2
[117] 2 2 3 2 3 2 3 2 2 2 3 3 2 2 2 2 2 3 2 2 2 2 3 2 2 2 3 2 2 2
[146] 2 3 2 2 3

Within cluster sum of squares by cluster:
[1] 15.15100 23.87947 39.82097
 (between_SS / total_SS =  88.4 %)

Available components:

[1] "cluster"      "centers"      "totss"        "withinss"
[5] "tot.withinss" "betweenss"    "size"         "iter"
[9] "ifault"
```

调用 table()函数显示聚类的结果，从以下结果可知，有 36 笔 versicolor 聚类至 virginica，有 48 笔 virginica 聚类至 versicolor：

```
# 混淆矩阵
> table(iris$Species, kmeans.result$cluster)

              1  2  3
  setosa     50  0  0
  versicolor  0  2 48
  virginica   0 36 14
```

以属性 Sepal.Length 为 X 轴、Sepal.Width 为 Y 轴，调用 plot()函数及 points()函数画出聚类簇的中心位置，如图 7-5 所示。

```
> plot(iris2[c("Sepal.Length", "Sepal.Width")], col = kmeans.result$cluster)
> points(kmeans.result$centers[,c("Sepal.Length", "Sepal.Width")], col = 1:3,
pch = 8, cex=2)
```

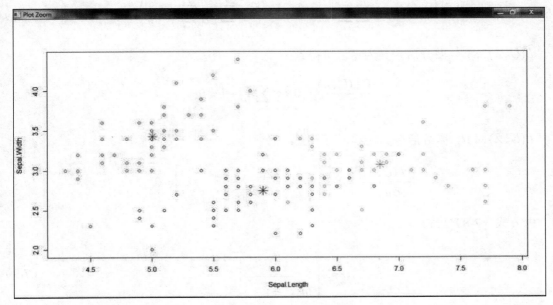

图 7-5　聚类簇的中心位置图

7.3　模糊 C 均值聚类算法

Bezdek 于 1973 年提出了模糊 C 均值聚类算法，它的目标函数应用了模糊理论的概念，使得每一输入向量不再仅归属于某一特定的聚类，而以其归属程度来表现隶属于各聚类的程度。

数据集 $X=\{x_1,x_2,\cdots,x_n\}$，模糊 C 均值聚类算法将数据集 X 分成 c 簇。令这 c 簇的簇心集为 $V=\{v_1,v_2,\cdots,v_c\}$，令数据点 x_i 对簇心 v_i 的隶属函数值为 u_{ij}，隶属矩阵 U 表示为：

$$U=\begin{bmatrix} u_{11} & u_{12} & \cdots & u_{1n} \\ u_{21} & u_{22} & \cdots & u_{2n} \\ \vdots & \vdots & \vdots & \vdots \\ u_{c1} & u_{c2} & \cdots & u_{cn} \end{bmatrix} \qquad (7\text{-}3)$$

簇心集 V 和数据点集 X 的误差为：

$$E(U,V:X)=\sum_{i=1}^{c}\sum_{j=1}^{n}(u_{ij})^m\|x_j-v_i\|^2 \qquad (7\text{-}4)$$

限制条件为:

$$\sum_{i=1}^{c} u_{ij} = 1 \tag{7-5}$$

依据 Lagrange 乘数方法,可得:

$$L(U,\lambda) = \sum_{j=1}^{n}\sum_{i=1}^{c}(u_{ij})^m \|x_j - v_i\|^2 - \sum_{j=1}^{n}\lambda_j\left(\sum_{i=1}^{c}u_{ij} - 1\right) \tag{7-6}$$

$L(U,\lambda)$ 对 λ_j 微分后令其为零,可得到:

$$\frac{\partial L(U,\lambda)}{\partial \lambda_j} = 0 \Leftrightarrow \sum_{i=1}^{c} u_{ij} - 1 = 0 \tag{7-7}$$

$L(U,\lambda)$ 对 u_{ij} 微分后令其为零,可得到:

$$\frac{\partial L(U,\lambda)}{\partial u_{ij}} = 0 \Leftrightarrow \left[m(u_{ij})^{m-1}\|x_j - v_i\|^2 - \lambda_j\right] = 0 \tag{7-8}$$

由公式(7-8)可得:

$$u_{ij} = \left(\frac{\lambda_j}{m\|x_j - v_i\|^2}\right)^{\frac{1}{m-1}} \tag{7-9}$$

由公式(7-7)和公式(7-9)可得:

$$\sum_{i=1}^{c} u_{ij} = \sum_{i=1}^{c}\left(\frac{\lambda_j}{m\|x_j - v_i\|^2}\right)^{\frac{1}{m-1}} = 1 \tag{7-10}$$

从公式(7-10)可得:

$$\left(\frac{\lambda_j}{m}\right)^{\frac{1}{m-1}} = 1 \bigg/ \sum_{i=1}^{c}\left(\frac{1}{m\|x_j - v_i\|^2}\right)^{\frac{1}{m-1}} \tag{7-11}$$

将公式(7-11)代入公式(7-9),可得:

$$u_{ij} = 1 \bigg/ \sum_{k=1}^{c}\left(\frac{\|x_j - v_i\|}{\|x_j - v_k\|}\right)^{\frac{2}{m-1}} \tag{7-12}$$

簇心 v_i 可调整为:

$$v_i = \frac{\sum_{j=1}^{n}(u_{ij})^m x_j - 1}{\sum_{j=1}^{n}(u_{ij})^m}, \quad 1 \leqslant i \leqslant c \qquad (7\text{-}13)$$

模糊 C 均值聚类算法依照下列 4 个步骤进行聚类：

（1）设置聚类数量 c、次方 m、误差容忍度 ε 和起始隶属矩阵 U_0。

（2）根据数据集和 U_0 计算出起始的聚类簇中心集。

（3）重新计算 u_{ij}，$1 \leqslant i \leqslant c$ 和 $1 \leqslant j \leqslant n$，修正各个聚类簇的中心值。

（4）计算出误差 $E = \sum_{i=1}^{c} v_i^{t+1} - v_i^{t}$，若 $E \leqslant \varepsilon$，则停止，否则回到步骤 3。

[程序范例 7-3]

首先引用 e1071 包和 iris 数据：

```
> library("e1071")
> data(iris)
```

调用 rbind()函数将 iris 数据赋值给 x 并调用 t()函数来进行矩阵的转置：

```
# 生成数据
> x<-rbind(iris$Sepal.Length, iris$Sepal.Width, iris$Petal.Length, iris$Petal.Width)
> x<-t(x)
```

调用 cmeans()函数，并设自变量聚类的簇数 centers=3、m=2、最大迭代次数 iter.max=5 及 verbose=TRUE 来显示运行过程的信息：

```
> result<-cmeans(x,m=2,centers=3,iter.max=500,verbose=TRUE,method="cmeans")
Iteration:    1, Error:  1.1065549797
Iteration:    2, Error:  0.5143839622
Iteration:    3, Error:  0.4317889832
Iteration:    4, Error:  0.4089850507
Iteration:    5, Error:  0.4047123528
Iteration:    6, Error:  0.4038350861
Iteration:    7, Error:  0.4035582273
Iteration:    8, Error:  0.4034493445
Iteration:    9, Error:  0.4034040642
Iteration:   10, Error:  0.4033850717
Iteration:   11, Error:  0.4033771134
Iteration:   12, Error:  0.4033737857
Iteration:   13, Error:  0.4033723967
Iteration:   14, Error:  0.4033718175
Iteration:   15, Error:  0.4033715763
Iteration:   16, Error:  0.4033714758
```

```
Iteration: 17, Error: 0.4033714340
Iteration: 18, Error: 0.4033714166
Iteration: 19, Error: 0.4033714094
Iteration: 20 converged, Error: 0.4033714063
> print(result)
Fuzzy c-means clustering with 3 clusters

Cluster centers:
       [,1]     [,2]     [,3]      [,4]
1  5.003966 3.414092 1.482811 0.2535443
2  5.888876 2.761049 4.363869 1.3972723
3  6.774943 3.052362 5.646696 2.0535137

Memberships:
              1            2            3
 [1,] 0.996623591 0.0023043863 0.0010720223
 [2,] 0.975851052 0.0166506280 0.0074983195
 [3,] 0.979824951 0.0137602302 0.0064148190
 [4,] 0.967425471 0.0224665169 0.0101080122
 [5,] 0.994470320 0.0037617495 0.0017679303
 [6,] 0.934570922 0.0448085703 0.0206205077
 [7,] 0.979490658 0.0140045709 0.0065047707
 [8,] 0.999547236 0.0003115527 0.0001412117
 [9,] 0.930376026 0.0477206324 0.0219033420
[10,] 0.982721838 0.0119363182 0.0053418439
[11,] 0.968040923 0.0217577753 0.0102013018
[12,] 0.992136600 0.0054321973 0.0024312023
[13,] 0.970638460 0.0201839493 0.0091775909
[14,] 0.922966474 0.0517966675 0.0252368585
[15,] 0.889755281 0.0726119584 0.0376327603
[16,] 0.841339798 0.1043526293 0.0543075725
[17,] 0.946923379 0.0355805082 0.0174961132
[18,] 0.996652682 0.0022885250 0.0010587927
[19,] 0.904130899 0.0655599590 0.0303091423
[20,] 0.979188234 0.0141579180 0.0066538478
[21,] 0.968602441 0.0218567895 0.0095407691
[22,] 0.984832257 0.0103737118 0.0047940313
[23,] 0.958656627 0.0275101878 0.0138331853
[24,] 0.979425664 0.0144624401 0.0061118961
[25,] 0.966916808 0.0232444617 0.0098387303
[26,] 0.973566137 0.0184599651 0.0079738984
[27,] 0.994844951 0.0035816529 0.0015733961
[28,] 0.993347956 0.0045652424 0.0020868017
[29,] 0.993675739 0.0043282354 0.0019960261
```

```
[30,]  0.979514890  0.0142022292  0.0062828813
[31,]  0.978726304  0.0147999074  0.0064737890
[32,]  0.974363071  0.0176971498  0.0079397789
[33,]  0.938518560  0.0411467611  0.0203346789
[34,]  0.904167145  0.0634837012  0.0323491536
[35,]  0.985064315  0.0103414539  0.0045942311
[36,]  0.984993039  0.0102016329  0.0048053285
[37,]  0.964183139  0.0242524395  0.0115644217
[38,]  0.990890771  0.0061826377  0.0029265911
[39,]  0.939681223  0.0410855538  0.0192332228
[40,]  0.998288945  0.0011777899  0.0005332648
[41,]  0.994727723  0.0035839504  0.0016883265
[42,]  0.850728207  0.1022757303  0.0469960626
[43,]  0.952613146  0.0321779226  0.0152089315
[44,]  0.979286701  0.0143835533  0.0063297452
[45,]  0.945268152  0.0381305220  0.0166013258
[46,]  0.972144726  0.0192234127  0.0086318609
[47,]  0.976792090  0.0158427458  0.0073651640
[48,]  0.974219574  0.0176533601  0.0081270657
[49,]  0.977219457  0.0155207770  0.0072597662
[50,]  0.997072388  0.0020082807  0.0009193315
[51,]  0.044575050  0.4542004553  0.5012244951
[52,]  0.029165723  0.7639574496  0.2068768275
[53,]  0.031265512  0.3686821493  0.6000523388
[54,]  0.049339121  0.8702294370  0.0804314421
[55,]  0.024110917  0.7586613416  0.2172277416
[56,]  0.005739272  0.9737967869  0.0204639407
[57,]  0.029798001  0.6729099761  0.2972920233
[58,]  0.285156106  0.5825391456  0.1323047481
[59,]  0.031234714  0.7209584447  0.2478068412
[60,]  0.074746602  0.8305638837  0.0946895143
[61,]  0.218400944  0.6365414420  0.1450576136
[62,]  0.009185489  0.9620737604  0.0287407510
[63,]  0.055621313  0.8432649512  0.1011137362
[64,]  0.012118006  0.8995880340  0.0882939601
[65,]  0.091629644  0.8161955414  0.0921748145
[66,]  0.041721090  0.6898262780  0.2684526317
[67,]  0.014153805  0.9331611163  0.0526850788
[68,]  0.025884884  0.9257567915  0.0483583250
[69,]  0.027145071  0.8354092081  0.1374457207
[70,]  0.051544584  0.8776981932  0.0707572230
[71,]  0.027618640  0.7215665254  0.2508148344
[72,]  0.019490218  0.9342822228  0.0462275596
[73,]  0.023972824  0.7053570878  0.2706700884
```

```
 [74,] 0.013936373 0.9027004542 0.0833631723
 [75,] 0.022907805 0.8758831346 0.1012090602
 [76,] 0.033942906 0.7547325843 0.2113245093
 [77,] 0.033606150 0.5235844624 0.4428093876
 [78,] 0.021184381 0.3062398204 0.6725757989
 [79,] 0.004939666 0.9688476662 0.0262126679
 [80,] 0.128301530 0.7670490155 0.1046494549
 [81,] 0.077790647 0.8323328142 0.0898765391
 [82,] 0.103762904 0.7955049785 0.1007321177
 [83,] 0.030971559 0.9187363048 0.0502921366
 [84,] 0.023972124 0.6562230039 0.3198048724
 [85,] 0.026379608 0.8911646942 0.0824556978
 [86,] 0.032144915 0.7971115685 0.1707435165
 [87,] 0.033454318 0.5553331286 0.4112125531
 [88,] 0.026864387 0.8576440693 0.1154915436
 [89,] 0.024084133 0.9289403177 0.0469755498
 [90,] 0.038167576 0.8994678822 0.0623645416
 [91,] 0.019624492 0.9311594581 0.0492160495
 [92,] 0.011621296 0.9156100918 0.0727686125
 [93,] 0.022508771 0.9357165622 0.0417746670
 [94,] 0.268911674 0.5983522463 0.1327360793
 [95,] 0.012673718 0.9588977271 0.0284285546
 [96,] 0.016778038 0.9455531803 0.0376687821
 [97,] 0.009572210 0.9674237953 0.0230039948
 [98,] 0.011408065 0.9438529899 0.0447389455
 [99,] 0.355330531 0.5201593350 0.1245101338
[100,] 0.012683753 0.9605895752 0.0267266715
[101,] 0.019358235 0.1207293282 0.8599124372
[102,] 0.029297301 0.6155055693 0.3551971299
[103,] 0.006075879 0.0381501798 0.9557739416
[104,] 0.012522179 0.1418956202 0.8455822010
[105,] 0.004753887 0.0376327136 0.9576133993
[106,] 0.035459060 0.1526643550 0.8118765853
[107,] 0.072969289 0.7599903494 0.1670403613
[108,] 0.021897303 0.1151298020 0.8629728954
[109,] 0.013991215 0.1173156540 0.8686931314
[110,] 0.024416681 0.1145852161 0.8609981025
[111,] 0.016760627 0.2098109300 0.7734284433
[112,] 0.015751889 0.2230559309 0.7611921800
[113,] 0.001165779 0.0100110315 0.9888231890
[114,] 0.034364279 0.6599077515 0.3057279699
[115,] 0.038368189 0.4609331558 0.5006986550
[116,] 0.014060308 0.1360268918 0.8499127998
[117,] 0.007158380 0.0795776206 0.9132639993
```

```
[118,] 0.050587296 0.1858085498 0.7636041543
[119,] 0.049181070 0.1932444361 0.7575744942
[120,] 0.032187261 0.7106728992 0.2571398394
[121,] 0.003828058 0.0256996694 0.9704722729
[122,] 0.033640470 0.7069207083 0.2594388215
[123,] 0.041969706 0.1748179146 0.7832123794
[124,] 0.022801193 0.5956354579 0.3815633491
[125,] 0.002913570 0.0217413242 0.9753451062
[126,] 0.012907281 0.0751612211 0.9119314979
[127,] 0.020979914 0.7076636051 0.2713564807
[128,] 0.023091768 0.6467848930 0.3301233394
[129,] 0.008355500 0.0829718259 0.9086726743
[130,] 0.014444647 0.0947351575 0.8908201958
[131,] 0.019785664 0.1074380046 0.8727763312
[132,] 0.050902247 0.1889836824 0.7601140702
[133,] 0.008954187 0.0848145950 0.9062312176
[134,] 0.023389370 0.5400881802 0.4365224498
[135,] 0.031182190 0.3926627258 0.5761550847
[136,] 0.028660979 0.1317658493 0.8395731717
[137,] 0.017224871 0.1286753609 0.8540997684
[138,] 0.009780293 0.1100245651 0.8801951417
[139,] 0.021725820 0.7498049032 0.2284692770
[140,] 0.003488313 0.0289513662 0.9675603210
[141,] 0.005076956 0.0376725566 0.9572504872
[142,] 0.015399238 0.1294491591 0.8551516025
[143,] 0.029297301 0.6155055693 0.3551971299
[144,] 0.005250021 0.0337345741 0.9610154048
[145,] 0.009701055 0.0631749934 0.9271239514
[146,] 0.011259697 0.1063482072 0.8823920961
[147,] 0.025795992 0.5074231362 0.4667808722
[148,] 0.012109692 0.1563603928 0.8315299155
[149,] 0.021579041 0.1889929058 0.7894280532
[150,] 0.026920096 0.5816808796 0.3913990239

Closest hard clustering:
  [1] 1 1 1 1 1 1 1 1 1 1 1 1 1 1 1 1 1 1 1 1 1 1 1 1 1 1 1 1 1 1 1 1 1 1 1 1
 [37] 1 1 1 1 1 1 1 1 1 1 1 1 1 1 1 3 2 3 2 2 2 2 2 2 2 2 2 2 2 2 2 2 2 2 2 2
 [73] 2 2 2 2 3 2 2 2 2 2 2 2 2 2 2 2 2 2 2 2 2 2 2 2 2 3 2 3 3 3 3 2 3
[109] 3 3 3 3 3 2 3 3 3 3 3 2 3 2 3 3 2 2 3 3 3 3 2 3 3 3 3 2 3 3 3 3 2 3
[145] 3 3 2 3 3 2

Available components:
[1] "centers"    "size"       "cluster"    "membership" "iter"
[6] "withinerror" "call"
```

调用 table()函数显示聚类的结果，由以下结果可知，有 13 笔 versicolor 聚类至 virginica，有 3 笔 virginica 聚类至 versicolor。

```
# 混淆矩阵
> table(iris$Species, result$cluster)

              1   2   3
  setosa     50   0   0
  versicolor  0  47   3
  virginica   0  13  37
```

7.4 聚类指标

由于使用非监督式的聚类算法时需要先决定聚类的数量，因此如何决定聚类的数量便是一项很重要的工作。聚类指标可评估聚类的效果，协助用户决定聚类的数量。

[程序范例 7-4]

首先引用 NbClust 包和 iris 数据：

```
> library(NbClust)
> data(iris)
```

由于是非监督式学习，因此不使用 iris 数据中的目标属性（Species）：

```
> data <-iris[,-c(5)]
```

调用 NbClust()函数，并设置自变量 distance="euclidean"、聚类数量介于 min.nc=2 到 max.nc=6 之间、method= "kmeans"及 index = "all"：

```
> NbClust(data, distance = "euclidean", min.nc=2, max.nc=6, method = "kmeans", index = "all")
*** : The Hubert index is a graphical method of determining the number of clusters.
In the plot of Hubert index, we seek a significant knee that corresponds to a
significant increase of the value of the measure i.e the significant peak in Hubert
index second differences plot.

*** : The D index is a graphical method of determining the number of clusters.
In the plot of D index, we seek a significant knee (the significant peak in Dindex
second differences plot) that corresponds to a significant increase of the value
of the measure.

All 150 observations were used.

*******************************************************************
```

```
* Among all indices:                                                
* 8 proposed 2 as the best number of clusters
* 11 proposed 3 as the best number of clusters
* 1 proposed 4 as the best number of clusters
* 2 proposed 5 as the best number of clusters
* 2 proposed 6 as the best number of clusters

                   ***** Conclusion *****                           

* According to the majority rule, the best number of clusters is  3


*******************************************************************
$All.index
       KL       CH Hartigan     CCC   Scott  Marriot    TrCovW   TraceW
2  5.9068 513.9245 137.9491 35.9428 1044.605 467371.6 1045.9696 152.3480
3 12.4890 561.6278  15.2384 37.6701 1246.668 273408.6  248.9814  78.8514
4  9.7643 530.7658  20.5286 36.4682 1359.280 229428.4  173.8973  57.2285
5  0.8290 495.5415  -3.5725 35.2758 1465.533 176536.0  117.4449  46.4462
6 17.4419 441.7370  16.2762 33.6090 1527.121 168608.2   75.9611  41.7044
   Friedman   Rubin  Cindex     DB Silhouette   Duda Pseudot2   Beale
2  732.8086  62.6152 0.2728 0.4744    0.6810 0.3153 186.7726  5.1885
3  801.6490 120.9780 0.3450 0.7256    0.5528 0.6779  27.5617  1.1284
4  874.3981 166.6878 0.3211 0.8436    0.4981 0.4309  63.3980  3.0858
5  989.3862 205.3837 0.2979 0.8571    0.4887 1.8245 -29.8255 -1.0623
6 1087.4777 228.7357 0.2850 0.9427    0.4841 2.3936 -19.7954 -1.3275
   Ratkowsky    Ball Ptbiserial    Frey McClain   Dunn Hubert SDindex Dindex
2     0.5462 76.1740     0.8345  1.7571  0.2723 0.0765 0.0019  0.9995 0.8556
3     0.4967 26.2838     0.7146 -20.7325 0.5255 0.0988 0.0021  1.5740 0.6480
4     0.4413 14.3071     0.6361  5.5984  0.7120 0.1365 0.0021  2.3932 0.5574
5     0.3997  9.2892     0.6148 -4.4141  0.7671 0.0823 0.0022  2.7979 0.5097
6     0.3682  6.9507     0.5679  0.8351  0.9164 0.1089 0.0022  4.3487 0.4783
    SDbw
2 0.1618
3 0.2257
4 0.3186
5 0.0538
6 0.0469

$All.CriticalValues
  CritValue_Duda CritValue_PseudoT2 Fvalue_Beale
2         0.6357            49.2854       0.0004
3         0.5842            41.2743       0.3437
4         0.4837            51.2421       0.0185
```

```
5         0.5173          61.5845       1.0000
6         0.3773          56.1121       1.0000
```

$Best.nc
```
                  KL       CH  Hartigan    CCC    Scott  Marriot   TrCovW
Number_clusters  6.0000   3.0000   3.0000   3.0000   3.0000     3.0    3.0000
Value_Index     17.4419 561.6278 122.7107  37.6701 202.0631 149982.9 796.9882
                 TraceW Friedman    Rubin  Cindex      DB Silhouette    Duda
Number_clusters  3.0000    5.000   5.0000  2.0000  2.0000      2.000  3.0000
Value_Index     51.8735  114.988 -15.3439  0.2728  0.4744      0.681  0.6779
                PseudoT2   Beale Ratkowsky    Ball PtBiserial    Frey McClain
Number_clusters  3.0000  3.0000   2.0000  3.0000    2.0000  2.0000  2.0000
Value_Index     27.5617  1.1284   0.5462 49.8902    0.8345  1.7571  0.2723
                  Dunn  Hubert SDindex  Dindex    SDbw
Number_clusters  4.0000       0  2.0000       0  6.0000
Value_Index      0.1365       0  0.9995       0  0.0469
```

$Best.partition
```
  [1] 1 1 1 1 1 1 1 1 1 1 1 1 1 1 1 1 1 1 1 1 1 1 1 1 1 1 1 1 1 1 1 1 1 1 1 1
 [37] 1 1 1 1 1 1 1 1 1 1 1 1 1 1 2 2 3 2 2 2 2 2 2 2 2 2 2 2 2 2 2 2 2 2 2 2
 [73] 2 2 2 2 3 2 2 2 2 2 2 2 2 2 2 2 2 2 2 2 2 2 2 2 2 2 3 2 3 3 3 3 2 3
[109] 3 3 3 3 2 2 3 3 3 3 3 2 3 2 3 3 3 2 3 3 3 3 3 3 3 2 3 3 3 3 3 3 3 2 3
[145] 3 3 2 3 3 2
```

以上数据显示，有 11 种聚类指标表示使用 3 簇是最好的聚类数量。

7.5 习题

调用 kmeans()函数将 insurance.csv 聚成 3 类并找出簇心。

第 8 章

演化式学习

演化式学习是指模拟自然界演化过程所建立的学习模型，通常将演化式学习所建立的模型通称为演化式算法。本章介绍常用的演化式算法，包含遗传算法及人工蜂群算法。

8.1 遗传算法

Holland 等人于 1975 年提出了遗传算法的基本理论，其基本精神在于模仿自然界中物竞天择、优胜劣汰的自然进化法则，选择物种对环境适应力较强的亲代（Parents Generation）并随机相互交换彼此的基因信息，以期产生更优秀的子代（Offspring Generation），经过筛选（Selection）后留下适应力最佳的物种，再继续交配、繁衍及筛选，如此重复，不断演化出对外在环境适应力最强的物种。

复制（Reproduction）、交配（Crossover）、突变（Mutation）是遗传算法基本的三种运行机制。遗传算法是将问题通过编码的方式转换到染色体（个体）结构上，然后利用适应函数（Fitness Function）的定义评估（Evaluation）解答的优劣程度，经由复制、交配、突变等运行机制产生新的染色体（Chromosome），并取代（Replace）在群族中表现不好的染色体。演化学习的过程是不断反复地评估、筛选，找到适应函数最佳的染色体以保留至下一个代（Generation），并继续演化，直到满足终止条件为止。遗传算法的基本概念如下：

（1）将问题以编码的方式对应到一个解（Solution），在遗传算法中称为染色体或个体（Individual）。

（2）利用并行搜索的概念，同时产生多组染色体来进行随机搜索，在遗传算法中称为族群（Population）。

（3）根据染色体的适应函数来评估解的优劣。

（4）通过演化复制的过程，利用随机的方式将上一代的部分基因移转到下一代身上，创造出不同于上一代的新染色体，这就是基因运行机制。

遗传算法的流程图如图 8-1 所示。

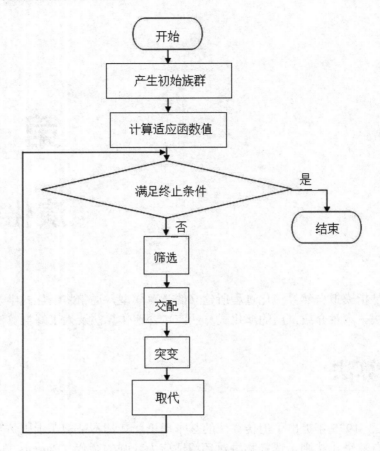

图 8-1　遗传算法的流程图

遗传算法求解的第一个步骤就是产生初始族群。使用二进制遗传算法时，可先将所要进行优化求解的问题参数编码成一定长度的二进制字符串，例如：

0110101101
1100011000

若解为实数，则可使用公式（8-1）将二进制数转换为实数：

$$x = B2D*(UB-LB)/(2^L-1)+LB \qquad (8-1)$$

其中：

- B2D：二进制数转换成十进制数。
- UB：上界。
- LB：下界。
- L：二进制数的字符串长度。
- x：输出的实数值。

遗传算法有三种主要的运算，分别为复制、交配及突变。首先，在进行复制运算前，必须先进行筛选。筛选的方法很多，例如轮盘赌选择法（Roulette Wheel Selection）、锦标赛选择法（Tournament Selection）。以轮盘赌选择法为例，它是根据每个染色体的适应函数值决定该染色体在轮盘上的面积，适应函数值越大，则面积越大，其被选到的概率越大，被选到的染色体被送到交配池中等待交配。锦标赛选择法是指，在每一代的演化过程中，随机选择两个或多个染色体，具有最大适应函数值的染色体即被选中送至交配池中。在进行交配的过程中，必须先决定交配概率的大小（Crossover Probability），每一对染色体都将按照交配概率来决定是否交配，希望经由交配过程能持续累积优秀的染色体，使得交配产生的子代适应函数值越来越好。常用的交配方法分别为单点交配（One-Point Crossover）、双点交配（Two-Point Crossover）、多点交配（Multi-point Crossover）、字罩交配（Mask Crossover）及概率均等交配（Uniform Crossover）等。单点交配就是在族群内随机选取的两段基因彼此交换，组成两个新的基因字符串。假设算法选择 A 的第 5 位为交配点：

```
A = 0110011111
B = 1100100000
```

交配点选择完毕后，接着将两个染色体中位于交配点后的所有位互换，结果如下：

```
A = 0110000000
B = 1100111111
```

交配完成后，A 演化成 0110000000，B 演化成 1100111111。经由复制与交配会产生许多不同的新染色体，单就以染色体上的基因来说，基因上未产生新的信息。因此，希望能通过某些方式使基因有新的信息产生，这种方法就是突变，如同先前的交配过程，突变过程也需要设置一个突变概率（Mutation Probability）。简单的突变运作方式是选取任一染色体中的某一基因，再将该点所代表的位取反，例如：

```
A = 010000000
```

假设算法选择 A 的第 8 位作为突变点，则：

```
A = 010000000 → A = 000000000
```

突变完成后，A 由 010000000 突变为 000000000。取代部分以精英政策为主，其目的是将旧族群中适应函数值最佳者与前几名的染色体存留下来，其余的以子代新染色体取代，成为新的族群。例如，当精英政策设置为 0.6 时，代表有 60%亲代的染色体保留下来（来自前一代族群中适应函数值的前 60%的染色体），而其余的 40%则用表现较佳的子代染色体来取代，成为新的族群。每一代都会重复上述动作，不断地更新族群中的染色体，通过精英政策的方式来留住表现较佳的染色体，直到满足终止条件为止。

[程序范例 8-1]

首先引用 GA 包：

```
> library("GA")
```

本范例中欲求 Max.$f(x) = 25 - x*x$，$-5 \leqslant x \leqslant 5$：

```
> f <- function(x)  25-x*x
> min <- -5
> max <- +5
```

调用 curve()函数绘出 $f(x)=25-x*x$，$-5\leqslant x\leqslant 5$，如图 8-2 所示。

```
# f(x)=25-x*x
> curve(f, min, max)
```

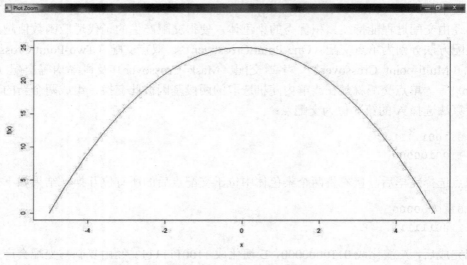

图 8-2 $f(x)=25-x*x$ 图

设置适应函数为 $f(x)$：

```
> fitness <- function(x) f(x)
```

调用 ga()函数，并设置自变量 popSize=50、pcrossover=0.8、pmutation=0.1、elitism=10、monitor=gaMonitor、maxiter=100，最优的解为 $f(x)=25$。

```
# 调用ga()函数，迭代数=100
> GA <- ga(type="real-valued",
+ fitness=fitness,
+ min=min,
+ max=max,
+ popSize = 50,
+ pcrossover = 0.8,
+ pmutation = 0.1,
+ elitism = 10,
+ monitor = gaMonitor,
+ maxiter = 100)

Iter = 1  | Mean = 16.75026 | Best = 24.99012
Iter = 2  | Mean = 22.87244 | Best = 24.99869
Iter = 3  | Mean = 23.82392 | Best = 24.99869
```

```
Iter = 4  | Mean = 23.77883 | Best = 24.99869
Iter = 5  | Mean = 23.25862 | Best = 24.99999
Iter = 6  | Mean = 23.89032 | Best = 24.99999
Iter = 7  | Mean = 24.16965 | Best = 24.99999
Iter = 8  | Mean = 24.31778 | Best = 25
Iter = 9  | Mean = 24.04458 | Best = 25
Iter = 10 | Mean = 24.19527 | Best = 25
Iter = 11 | Mean = 24.66865 | Best = 25
Iter = 12 | Mean = 24.76368 | Best = 25
Iter = 13 | Mean = 23.61497 | Best = 25
Iter = 14 | Mean = 24.51597 | Best = 25
Iter = 15 | Mean = 24.84668 | Best = 25
Iter = 16 | Mean = 24.70468 | Best = 25
Iter = 17 | Mean = 23.72271 | Best = 25
Iter = 18 | Mean = 23.83589 | Best = 25
Iter = 19 | Mean = 23.83062 | Best = 25
Iter = 20 | Mean = 24.88215 | Best = 25
Iter = 21 | Mean = 23.84913 | Best = 25
Iter = 22 | Mean = 24.12524 | Best = 25
Iter = 23 | Mean = 24.00631 | Best = 25
Iter = 24 | Mean = 24.24212 | Best = 25
Iter = 25 | Mean = 24.33541 | Best = 25
Iter = 26 | Mean = 24.6339  | Best = 25
Iter = 27 | Mean = 24.40812 | Best = 25
Iter = 28 | Mean = 24.80727 | Best = 25
Iter = 29 | Mean = 23.39615 | Best = 25
Iter = 30 | Mean = 24.64121 | Best = 25
Iter = 31 | Mean = 23.97989 | Best = 25
Iter = 32 | Mean = 23.76933 | Best = 25
Iter = 33 | Mean = 23.93873 | Best = 25
Iter = 34 | Mean = 23.98081 | Best = 25
Iter = 35 | Mean = 24.16118 | Best = 25
Iter = 36 | Mean = 24.31206 | Best = 25
Iter = 37 | Mean = 23.61612 | Best = 25
Iter = 38 | Mean = 24.56559 | Best = 25
Iter = 39 | Mean = 24.53877 | Best = 25
Iter = 40 | Mean = 23.64468 | Best = 25
Iter = 41 | Mean = 24.3073  | Best = 25
Iter = 42 | Mean = 24.52696 | Best = 25
Iter = 43 | Mean = 24.73332 | Best = 25
Iter = 44 | Mean = 23.64634 | Best = 25
Iter = 45 | Mean = 24.46134 | Best = 25
Iter = 46 | Mean = 24.07619 | Best = 25
Iter = 47 | Mean = 23.87074 | Best = 25
```

```
Iter = 48  | Mean = 24.18712 | Best = 25
Iter = 49  | Mean = 24.33388 | Best = 25
Iter = 50  | Mean = 24.12963 | Best = 25
Iter = 51  | Mean = 24.66332 | Best = 25
Iter = 52  | Mean = 24.45472 | Best = 25
Iter = 53  | Mean = 23.33875 | Best = 25
Iter = 54  | Mean = 24.32067 | Best = 25
Iter = 55  | Mean = 24.01558 | Best = 25
Iter = 56  | Mean = 23.85615 | Best = 25
Iter = 57  | Mean = 24.2742  | Best = 25
Iter = 58  | Mean = 24.4183  | Best = 25
Iter = 59  | Mean = 23.67884 | Best = 25
Iter = 60  | Mean = 23.69734 | Best = 25
Iter = 61  | Mean = 23.4972  | Best = 25
Iter = 62  | Mean = 24.9315  | Best = 25
Iter = 63  | Mean = 24.43032 | Best = 25
Iter = 64  | Mean = 22.74806 | Best = 25
Iter = 65  | Mean = 24.14798 | Best = 25
Iter = 66  | Mean = 23.36102 | Best = 25
Iter = 67  | Mean = 24.2989  | Best = 25
Iter = 68  | Mean = 24.67911 | Best = 25
Iter = 69  | Mean = 23.88688 | Best = 25
Iter = 70  | Mean = 24.46076 | Best = 25
Iter = 71  | Mean = 24.44833 | Best = 25
Iter = 72  | Mean = 24.1129  | Best = 25
Iter = 73  | Mean = 23.45282 | Best = 25
Iter = 74  | Mean = 24.12345 | Best = 25
Iter = 75  | Mean = 23.4959  | Best = 25
Iter = 76  | Mean = 23.90719 | Best = 25
Iter = 77  | Mean = 24.46473 | Best = 25
Iter = 78  | Mean = 24.51689 | Best = 25
Iter = 79  | Mean = 23.56051 | Best = 25
Iter = 80  | Mean = 23.99792 | Best = 25
Iter = 81  | Mean = 24.5327  | Best = 25
Iter = 82  | Mean = 22.67488 | Best = 25
Iter = 83  | Mean = 24.0043  | Best = 25
Iter = 84  | Mean = 23.64904 | Best = 25
Iter = 85  | Mean = 24.47736 | Best = 25
Iter = 86  | Mean = 23.67569 | Best = 25
Iter = 87  | Mean = 23.31366 | Best = 25
Iter = 88  | Mean = 23.93136 | Best = 25
Iter = 89  | Mean = 23.81057 | Best = 25
Iter = 90  | Mean = 23.2838  | Best = 25
Iter = 91  | Mean = 24.3743  | Best = 25
```

```
Iter = 92  | Mean = 24.27372 | Best = 25
Iter = 93  | Mean = 22.63764 | Best = 25
Iter = 94  | Mean = 23.6324  | Best = 25
Iter = 95  | Mean = 24.77453 | Best = 25
Iter = 96  | Mean = 24.48018 | Best = 25
Iter = 97  | Mean = 23.78382 | Best = 25
Iter = 98  | Mean = 23.28074 | Best = 25
Iter = 99  | Mean = 23.29128 | Best = 25
Iter = 100 | Mean = 24.03805 | Best = 25
```

调用 plot() 函数画出运行的结果，如图 8-3 所示。

```
> plot(GA)
```

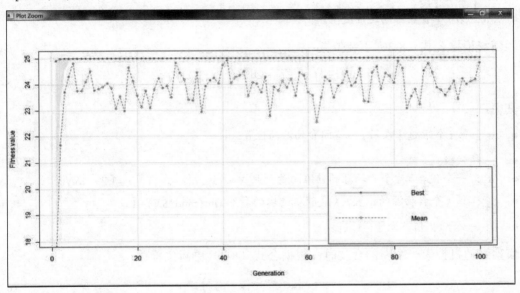

图 8-3　遗传算法运行的结果

8.2　人工蜂群算法

人工蜂群算法是建立在蜜蜂群体智慧的基础上，其核心主要由觅食蜂（The Employed Bee）、跟随蜂（The Onlooker Bee）、侦察蜂（The Scout）以及食物源（Sources）组成。蜜蜂对食物源的搜索有三个步骤：（1）觅食蜂发现食物源并记录下花蜜的数量；（2）跟随蜂根据觅食蜂所提供的花蜜信息来选定到哪个食物源采蜜；（3）侦查蜂随机搜索蜂巢附近的食物源，以寻找新的食物源。食物源相当于优化问题中解的位置，在人工蜂群算法中，食物源（解）的价值由适应度（Fitness）值来表示。某个解经过有限次循环后没有得到改善，代表其陷入局部最优解，那么这个食物源就应该被放弃，放弃发现食物源的觅食蜂将成为侦察蜂，并随机搜索一个新的食物。

Karaboga 提出了人工蜂群（Artificial Bee Colony，ABC）算法并成功地将其应用于函数数值优化问题。在人工蜂群算法中，人工蜂群一半由觅食蜂构成，另一半由跟随蜂构成。每一处食物源仅仅有一只觅食蜂，也就是说食物源和觅食蜂的数量相等。觅食蜂存储了某一个食物源的相关数据（位置及花蜜数量等），并将这些信息以一定的概率与其他人工蜂分享。觅食蜂在蜂巢内将它们的信息通过舞蹈传递给跟随蜂，跟随蜂随机选择一个食物源并变为觅食蜂。在人工蜂群算法中，跟随蜂选择食物源的概率（P_i）根据公式（8-2）计算：

$$P_i = \frac{\text{Fit}_i}{\sum_{i=1}^{SN}\text{Fit}_i} \quad (8\text{-}2)$$

其中：

- SN：食物源的数量。
- Fit_i：第 i 个食物源的适应度值。

跟随蜂根据公式（8-3）更新位置：

$$v_{ij}(t+1) = x_{ij}(t) + \varnothing\left(x_{ij}(t) - x_{kj}(t)\right) \quad (8\text{-}3)$$

其中：

- v_{ij}：第 i 个跟随蜂在第 j 维空间食物源的位置。
- t：迭代数。
- X_i：一个代表第 i 个食物源的 d 维向量，且 $X_i = (x_{i1}, x_{i2}, \cdots, x_{id})$，$i = 1, 2, \cdots, SN$。
- x_{kj}：第 k 个食物源的位置，k 是随机选取的，$k = \text{int}(\text{rand}*SN) + 1$。
- $\varnothing$：一个随机值，介于[-1, 1]。

侦察蜂随机搜索一个新的食物时，根据公式（8-4）更新位置：

$$x_{ij} = x_{\min}^j + \gamma * \left(x_{\max}^j - x_{\min}^j\right) \quad (8\text{-}4)$$

其中：

- γ：一个介于[0,1]的随机值。
- $x_{\min}^j$：位于 j 维空间食物源位置的下界。
- $x_{\max}^j$：位于 j 维空间食物源位置的上界。

[程序范例 8-2]

首先引用 ABCoptim 包：

```
> library(ABCoptim)
```

本范例中欲求 $\text{Min}.f(x) = -\cos x_1 * \cos x_2 * \exp\left(-\left((x_1-\pi)^2 + (x_2-\pi)^2\right)\right)$：

```
#f(x)=-cosx₁*cosx₂*exp(-((x₁-π)²+(x₂-π)²))
> fun <- function(x) {
```

```
+     -cos(x[1])*cos(x[2])*exp(-((x[1] - pi)^2 + (x[2] - pi)^2))
+ }
```

调用 abc_optim()函数并设置自变量，调用 rep(0, 2)函数产生初始值为(0, 0)，解的范围为 $-20 \leqslant x_1, x_2 \leqslant 20$，迭代数 criter=1000：

```
# 调用abc_optim()函数
> abc_optim(rep(0,2), fun, lb=-20, ub=20, criter=1000)
$par
[1] 3.141593 3.141593

$value
[1] -1

$counts
function
    1000
```

由以上结果可知，x_1, x_2=3.141593 时 $f(x)$=-1。

[程序范例8-3]

首先引用 ABCoptim 包：

```
> library(ABCoptim)
```

本范例中欲求 $Min. f(x) = 10*\sin(0.3*x)*\sin(1.3*x^2) + 0.00001*x^4 + 0.2*x + 80$：

```
# 设置 f(x)
> fw <- function (x)
+     10*sin(0.3*x)*sin(1.3*x^2) + 0.00001*x^4 + 0.2*x+80
```

调用 abc_optim()函数并设置自变量，产生初始值为 50，解的范围为 $-100 \leqslant x_1, x_2 \leqslant 100$，迭代数 criter=1000：

```
# 调用abc_optim()函数
> abc_optim(50, fw, lb=-100, ub=100, criter=1000)
$par
[1] -15.81515

$value
[1] 67.46773

$counts
function
    1000
```

由以上结果可知，x= -15.81515 时 $f(x)$=67.46773。

第 9 章

混合式学习

混合式学习是使用两种以上学习法的优点，借以提升单一学习法的性能或效率。

9.1 人工蜂群算法混合决策树

使用 C 5.0 决策树时，最少案例数量（minCases）、剪枝树置信水平（Confidence Level，CF）这两项参数在面对各种不同问题时会有不同的最佳参数组合，因此可使用人工蜂群算法来调整决策树参数，以求得较好的分类准确率。

[程序范例 9-1]

首先引用 C50 和 ABCoptim 包：

```
> library(C50)
> library(ABCoptim)
```

建立自定义函数 test.error()，其输入自变量 xx[1]为 CF 值（1≤CF≤100）、xx[2]为 minCases 值（minCases≥2），输出值为测试数据的错误率。本范例中以 best_CF、best_minCases 及 min_error 来记录最佳的 CF、minCases 及测试数据的最小错误率。注意，必须使用"<-"才能改变自定义函数外部对象的值。

```
# 自定义函数
> test.error <- function(xx){
+   c <- C5.0Control(subset = FALSE,
+             bands = 0,
+             winnow = FALSE,
+             noGlobalPruning = FALSE,
+             CF = xx[1]/100,
```

```r
+                         minCases = floor(xx[2]),
+                         fuzzyThreshold = FALSE,
+                         sample = 0,          # for holdout
+                         seed = sample.int(4096, size = 1) - 1L,
+                         earlyStopping = TRUE
+   )
+
+   treeModel <- C5.0(x = iris.train[, -5], y = iris.train$Species,control =c)
+   summary(treeModel)
+
+   test.output <- predict(treeModel, iris.test[, -5], type = "class")
+
+   n=length(test.output)
+   number=0
+   for( i in 1:n)
+   {
+     if (test.output[i] != iris.test[i,5])
+     {
+       number=number+1              # Error
+     }
+   }
+
+   error_value <- number/n*100
+
+   if (start_index == TRUE)
+   {
+     best_CF <<- c$CF          # Keep best global parameters
+     best_minCases <<- c$minCases
+     min_error <<- error_value
+   }
+
+   if (error_value < min_error )
+   {
+     best_CF <<- c$CF          # Keep best global parameters
+     best_minCases <<- c$minCases
+     min_error <<- error_value
+   }
+
+   error=error_value
+   return(error)
+ }
```

在主程序中，使用 iris 数据：

```
>##############################################################
> #  Main program
>##############################################################
> data(iris)
```

使用 50%作为测试数据：

```
> np = ceiling(0.5*nrow(iris))
> np
[1] 75
```

区分测试数据 iris.test 和训练数据 iris.train：

```
> iris.test = iris[1:np,]
> iris.train = iris[np+1:nrow(iris),]
```

先以初始值 CF=85%、minCases=25 运行 C5.0 决策树，并返回分类错误率为 68%：

```
# 设置全局变量(Global value)
> start_index <<- TRUE
> xx=c(85,2)                          # 初始值
> test.error(xx)
[1] 68
```

接着调用 abc_optim()函数，使用相同初始值，解的范围 2≤CF，minCases≤100，最大迭代数 maxCycle=10。运行后分类错误率从 68%降至 66.66667%。

```
> start_index <<- FALSE      # Global value
> abc_optim(xx,test.error,lb=2, ub=100, maxCycle=10)
$par
[1] 32.26316 32.26316

$value
[1] 66.66667

$counts
function
     10
> accuracy=100-min_error
> accuracy
[1] 33.33333
```

9.2 遗传算法混合人工神经网络

我们也可以使用遗传算法来调整人工神经网络参数，以得到更好的效果。ANN 包中提供了 ANNGA() 函数及演示范例。

[程序范例 9-2]

```
> data("dataANN")

> 调用ANNGA()函数，maxGen =100
> ANNGA(x =input,
+ y =output,
+ design =c(1, 3, 1),
+ population =100,
+ mutation = 0.3,
+ crossover = 0.7,
+ maxGen =100,
+ error =0.001)

***cycle***
Generation: 1  Best population fitness : 0.05981754 Mean of population:0.25887034
  Best chromosome->-20.36/13.89/1.50/-23.98/-9.71/9.35/15.48/-14.58/0.68/-15.66/

***cycle***
Generation: 2  Best population fitness : 0.03989556 Mean of population: 0.23688910
  Best chromosome->32.96/18.16/-2.28/-30.37/3.01/-16.76/0.88/21.82/13.42/-1.07/

***cycle***
Generation: 3  Best population fitness : 0.03989556 Mean of population: 0.22232952
  Best chromosome->32.96/18.16/-2.28/-30.37/3.01/-16.76/0.88/21.82/13.42/-1.07/

***cycle***
Generation: 4  Best population fitness : 0.03880315 Mean of population: 0.20310358
  Best chromosome->-0.51/15.81/18.32/15.46/-16.12/-6.78/3.00/8.42/35.82/-11.58/
```

```
***cycle***
Generation:  5  Best population fitness : 0.03880315 Mean of population: 0.18924708
    Best chromosome->-0.51/15.81/18.32/15.46/-16.12/-6.78/3.00/8.42/35.82/-11.58/

***cycle***
Generation:  6  Best population fitness : 0.03789907 Mean of population: 0.17406013
    Best chromosome->-23.55/-25.39/8.95/14.06/19.25/11.80/-9.81/-2.75/17.90/-15.04/

***cycle***
Generation:  7  Best population fitness : 0.03789907 Mean of population: 0.16399078
    Best chromosome->-23.55/-25.39/8.95/14.06/19.25/11.80/-9.81/-2.75/17.90/-15.04/

***cycle***
Generation:  8  Best population fitness : 0.03789907 Mean of population: 0.15730231
    Best chromosome->-23.55/-25.39/8.95/14.06/19.25/11.80/-9.81/-2.75/17.90/-15.04/

***cycle***
Generation:  9  Best population fitness : 0.03789907 Mean of population: 0.14605940
    Best chromosome->-23.55/-25.39/8.95/14.06/19.25/11.80/-9.81/-2.75/17.90/-15.04/

***cycle***
Generation:  10  Best population fitness : 0.03789907 Mean of population: 0.13908024
    Best chromosome->-23.55/-25.39/8.95/14.06/19.25/11.80/-9.81/-2.75/17.90/-15.04/

***cycle***
Generation:  11  Best population fitness : 0.03789907 Mean of population: 0.13207254
    Best chromosome->-23.55/-25.39/8.95/14.06/19.25/11.80/-9.81/-2.75/17.90/-15.04/

***cycle***
```

Generation: 12 Best population fitness : 0.03789907 Mean of population: 0.12698297
 Best chromosome->-23.55/-25.39/8.95/14.06/19.25/11.80/-9.81/-2.75/17.90/-15.04/

 cycle
 Generation: 13 Best population fitness : 0.03789907 Mean of population: 0.12094905
 Best chromosome->-23.55/-25.39/8.95/14.06/19.25/11.80/-9.81/-2.75/17.90/-15.04/

 cycle
 Generation: 14 Best population fitness : 0.03789907 Mean of population: 0.11631263
 Best chromosome->-23.55/-25.39/8.95/14.06/19.25/11.80/-9.81/-2.75/17.90/-15.04/

 cycle
 Generation: 15 Best population fitness : 0.03697816 Mean of population: 0.11132284
 Best chromosome->14.83/-15.09/34.34/8.66/-7.20/-20.87/-4.18/0.87/-5.00/-0.78/

 cycle
 Generation: 16 Best population fitness : 0.03647556 Mean of population: 0.10676000
 Best chromosome->14.35/-15.80/-1.29/-13.05/10.05/-20.28/-4.23/21.93/-0.11/-0.03/

 cycle
 Generation: 17 Best population fitness : 0.03647556 Mean of population: 0.10152121
 Best chromosome->14.35/-15.80/-1.29/-13.05/10.05/-20.28/-4.23/21.93/-0.11/-0.03/

 cycle
 Generation: 18 Best population fitness : 0.03647556 Mean of population: 0.09759118
 Best chromosome->14.35/-15.80/-1.29/-13.05/10.05/-20.28/-4.23/21.93/-0.11/-0.03/

 cycle
 Generation: 19 Best population fitness : 0.03647556 Mean of population: 0.09328037

Best chromosome->14.35/-15.80/-1.29/-13.05/10.05/-20.28/-4.23/21.93/-0.11/-0.03/

   cycle
   Generation: 20 Best population fitness : 0.03647556 Mean of population: 0.09093867
   Best chromosome->14.35/-15.80/-1.29/-13.05/10.05/-20.28/-4.23/21.93/-0.11/-0.03/

   cycle
   Generation: 21 Best population fitness : 0.03647556 Mean of population: 0.08929980
   Best chromosome->14.35/-15.80/-1.29/-13.05/10.05/-20.28/-4.23/21.93/-0.11/-0.03/

   cycle
   Generation: 22 Best population fitness : 0.03647556 Mean of population: 0.08791822
   Best chromosome->14.35/-15.80/-1.29/-13.05/10.05/-20.28/-4.23/21.93/-0.11/-0.03/

   cycle
   Generation: 23 Best population fitness : 0.03647556 Mean of population: 0.08586487
   Best chromosome->14.35/-15.80/-1.29/-13.05/10.05/-20.28/-4.23/21.93/-0.11/-0.03/

   cycle
   Generation: 24 Best population fitness : 0.03647556 Mean of population: 0.08391625
   Best chromosome->14.35/-15.80/-1.29/-13.05/10.05/-20.28/-4.23/21.93/-0.11/-0.03/

   cycle
   Generation: 25 Best population fitness : 0.03538457 Mean of population: 0.08210314
   Best chromosome->14.83/-23.51/3.32/8.66/12.33/-13.19/-4.18/0.87/-5.00/-0.78/

   cycle
   Generation: 26 Best population fitness : 0.03538457 Mean of population: 0.07831619
   Best chromosome->14.83/-23.51/3.32/8.66/12.33/-13.19/-4.18/0.87/-5.00/-0.78/

```
    ***cycle***
    Generation:  27  Best population fitness : 0.03538457 Mean of population:
0.07424434
    Best chromosome->14.83/-23.51/3.32/8.66/12.33/-13.19/-4.18/0.87/-5.00/
-0.78/

    ***cycle***
    Generation:  28  Best population fitness : 0.03538457 Mean of population:
0.07256437
    Best chromosome->14.83/-23.51/3.32/8.66/12.33/-13.19/-4.18/0.87/-5.00/
-0.78/

    ***cycle***
    Generation:  29  Best population fitness : 0.03538457 Mean of population:
0.07032876
    Best chromosome->14.83/-23.51/3.32/8.66/12.33/-13.19/-4.18/0.87/-5.00/
-0.78/

    ***cycle***
    Generation:  30  Best population fitness : 0.03538457 Mean of population:
0.06901380
    Best chromosome->14.83/-23.51/3.32/8.66/12.33/-13.19/-4.18/0.87/-5.00/
-0.78/

    ***cycle***
    Generation:  31  Best population fitness : 0.03538457 Mean of population:
0.06769783
    Best chromosome->14.83/-23.51/3.32/8.66/12.33/-13.19/-4.18/0.87/-5.00/
-0.78/

    ***cycle***
    Generation:  32  Best population fitness : 0.03538457 Mean of population:
0.06589963
    Best chromosome->14.83/-23.51/3.32/8.66/12.33/-13.19/-4.18/0.87/-5.00/
-0.78/

    ***cycle***
    Generation:  33  Best population fitness : 0.03538457 Mean of population:
0.06435588
    Best chromosome->14.83/-23.51/3.32/8.66/12.33/-13.19/-4.18/0.87/-5.00/
-0.78/

    ***cycle***
```

Generation: 34 Best population fitness : 0.03538457 Mean of population: 0.06406965
     Best chromosome->14.83/-23.51/3.32/8.66/12.33/-13.19/-4.18/0.87/-5.00/-0.78/

     cycle
   Generation: 35 Best population fitness : 0.03538457 Mean of population: 0.06037107
     Best chromosome->14.83/-23.51/3.32/8.66/12.33/-13.19/-4.18/0.87/-5.00/-0.78/

     cycle
   Generation: 36 Best population fitness : 0.03538457 Mean of population: 0.05790038
     Best chromosome->14.83/-23.51/3.32/8.66/12.33/-13.19/-4.18/0.87/-5.00/-0.78/

     cycle
   Generation: 37 Best population fitness : 0.03538457 Mean of population: 0.05684195
     Best chromosome->14.83/-23.51/3.32/8.66/12.33/-13.19/-4.18/0.87/-5.00/-0.78/

     cycle
   Generation: 38 Best population fitness : 0.03538457 Mean of population: 0.05615396
     Best chromosome->14.83/-23.51/3.32/8.66/12.33/-13.19/-4.18/0.87/-5.00/-0.78/

     cycle
   Generation: 39 Best population fitness : 0.03538457 Mean of population: 0.05344059
     Best chromosome->14.83/-23.51/3.32/8.66/12.33/-13.19/-4.18/0.87/-5.00/-0.78/

     cycle
   Generation: 40 Best population fitness : 0.03538457 Mean of population: 0.05124761
     Best chromosome->14.83/-23.51/3.32/8.66/12.33/-13.19/-4.18/0.87/-5.00/-0.78/

     cycle
   Generation: 41 Best population fitness : 0.03538457 Mean of population: 0.05065232

Best chromosome->14.83/-23.51/3.32/8.66/12.33/-13.19/-4.18/0.87/-5.00/-0.78/

   cycle
   Generation: 42 Best population fitness : 0.03431974 Mean of population: 0.04800988
    Best chromosome->17.69/-13.49/34.59/-61.31/-2.44/-23.15/-0.19/2.56/16.33/0.09/

   cycle
   Generation: 43 Best population fitness : 0.03431974 Mean of population: 0.04658839
    Best chromosome->17.69/-13.49/34.59/-61.31/-2.44/-23.15/-0.19/2.56/16.33/0.09/

   cycle
   Generation: 44 Best population fitness : 0.03431974 Mean of population: 0.04567413
    Best chromosome->17.69/-13.49/34.59/-61.31/-2.44/-23.15/-0.19/2.56/16.33/0.09/

   cycle
   Generation: 45 Best population fitness : 0.03350304 Mean of population: 0.04370310
    Best chromosome->10.08/-13.20/-51.64/-43.34/14.68/-18.22/-52.58/8.05/28.31/0.14/

   cycle
   Generation: 46 Best population fitness : 0.03350304 Mean of population: 0.04300280
    Best chromosome->10.08/-13.20/-51.64/-43.34/14.68/-18.22/-52.58/8.05/28.31/0.14/

   cycle
   Generation: 47 Best population fitness : 0.03350304 Mean of population: 0.04107488
    Best chromosome->10.08/-13.20/-51.64/-43.34/14.68/-18.22/-52.58/8.05/28.31/0.14/

   cycle
   Generation: 48 Best population fitness : 0.03350304 Mean of population: 0.04033673
    Best chromosome->10.08/-13.20/-51.64/-43.34/14.68/-18.22/-52.58/8.05/28.31/0.14/

```
***cycle***
Generation:  49  Best population fitness : 0.03350304 Mean of population:
0.03908786
  Best chromosome->10.08/-13.20/-51.64/-43.34/14.68/-18.22/-52.58/8.05/28.31/
0.14/

***cycle***
Generation:  50  Best population fitness : 0.03350304 Mean of population:
0.03899032
  Best chromosome->10.08/-13.20/-51.64/-43.34/14.68/-18.22/-52.58/8.05/28.31/
0.14/

***cycle***
Generation:  51  Best population fitness : 0.03350304 Mean of population:
0.03897411
  Best chromosome->10.08/-13.20/-51.64/-43.34/14.68/-18.22/-52.58/8.05/28.31/0.14/

***cycle***
Generation:  52  Best population fitness : 0.03350304 Mean of population:
0.03888024
  Best chromosome->10.08/-13.20/-51.64/-43.34/14.68/-18.22/-52.58/8.05/28.31/0.14/

***cycle***
Generation:  53  Best population fitness : 0.03350304 Mean of population:
0.03876724
  Best chromosome->10.08/-13.20/-51.64/-43.34/14.68/-18.22/-52.58/8.05/28.31/0.14/

***cycle***
Generation:  54  Best population fitness : 0.03350304 Mean of population:
0.03831313
  Best chromosome->10.08/-13.20/-51.64/-43.34/14.68/-18.22/-52.58/8.05/28.31/0.14/

***cycle***
Generation:  55  Best population fitness : 0.03341240 Mean of population:
0.03777473
  Best chromosome->13.88/-16.12/14.95/-32.53/-6.31/-3.61/-16.03/30.22/19.17/0.03/

***cycle***
Generation:  56  Best population fitness : 0.03341240 Mean of population:
0.03764631
  Best chromosome->13.88/-16.12/14.95/-32.53/-6.31/-3.61/-16.03/30.22/19.17/0.03/

***cycle***
```

Generation: 57 Best population fitness : 0.03227431 Mean of population: 0.03724999
　　Best chromosome->10.90/-42.97/-30.51/-65.17/3.04/-7.17/-48.00/8.93/-25.48/0.10/

　　cycle
　Generation: 58 Best population fitness : 0.03227431 Mean of population: 0.03720404
　　Best chromosome->10.90/-42.97/-30.51/-65.17/3.04/-7.17/-48.00/8.93/-25.48/0.10/

　　cycle
　Generation: 59 Best population fitness : 0.03227431 Mean of population: 0.03711149
　　Best chromosome->10.90/-42.97/-30.51/-65.17/3.04/-7.17/-48.00/8.93/-25.48/0.10/

　　cycle
　Generation: 60 Best population fitness : 0.03227431 Mean of population: 0.03647350
　　Best chromosome->10.90/-42.97/-30.51/-65.17/3.04/-7.17/-48.00/8.93/-25.48/0.10/

　　cycle
　Generation: 61 Best population fitness : 0.03177191 Mean of population:0.03634610
　　Best chromosome->16.25/-12.34/16.80/-46.88/-30.21/-22.08/-1.62/19.55/10.75/0.09/

　　cycle
　Generation: 62 Best population fitness : 0.02853393 Mean of population: 0.03619104
　　Best chromosome->-20.36/-21.39/23.15/-39.82/9.42/-4.25/-4.70/0.95/-0.41/0.11/

　　cycle
　Generation: 63 Best population fitness : 0.02428789 Mean of population: 0.03595940
　　Best chromosome->10.70/-19.86/36.83/-20.32/18.88/-39.12/-40.71/-0.97/-23.64/0.08/

　　cycle
　Generation: 64 Best population fitness : 0.02428789 Mean of population: 0.03590585
　　Best chromosome->10.70/-19.86/36.83/-20.32/18.88/-39.12/-40.71/-0.97/-23.64/0.08/

　　cycle
　Generation: 65 Best population fitness : 0.02428789 Mean of population: 0.03580409

Best chromosome->10.70/-19.86/36.83/-20.32/18.88/-39.12/-40.71/-0.97/-23.64/0.08/

   cycle
   Generation: 66 Best population fitness : 0.02428789 Mean of population: 0.03565368
   Best chromosome->10.70/-19.86/36.83/-20.32/18.88/-39.12/-40.71/-0.97/-23.64/0.08/

   cycle
   Generation: 67 Best population fitness : 0.02428789 Mean of population: 0.03563665
   Best chromosome->10.70/-19.86/36.83/-20.32/18.88/-39.12/-40.71/-0.97/-23.64/0.08/

   cycle
   Generation: 68 Best population fitness : 0.02428789 Mean of population: 0.03554259
   Best chromosome->10.70/-19.86/36.83/-20.32/18.88/-39.12/-40.71/-0.97/-23.64/0.08/

   cycle
   Generation: 69 Best population fitness : 0.02428789 Mean of population: 0.03540210
   Best chromosome->10.70/-19.86/36.83/-20.32/18.88/-39.12/-40.71/-0.97/-23.64/0.08/

   cycle
   Generation: 70 Best population fitness : 0.02428789 Mean of population: 0.03533953
   Best chromosome->10.70/-19.86/36.83/-20.32/18.88/-39.12/-40.71/-0.97/-23.64/0.08/

   cycle
   Generation: 71 Best population fitness : 0.02428789 Mean of population: 0.03521669
   Best chromosome->10.70/-19.86/36.83/-20.32/18.88/-39.12/-40.71/-0.97/-23.64/0.08/

   cycle
   Generation: 72 Best population fitness : 0.02428789 Mean of population: 0.03497467
   Best chromosome->10.70/-19.86/36.83/-20.32/18.88/-39.12/-40.71/-0.97/-23.64/0.08/

```
***cycle***
   Generation:  73  Best population fitness : 0.02428789 Mean of population:
0.03496585
    Best chromosome->10.70/-19.86/36.83/-20.32/18.88/-39.12/-40.71/-0.97/
-23.64/0.08/

***cycle***
   Generation:  74  Best population fitness : 0.02428789 Mean of population:
0.03482951
    Best chromosome->10.70/-19.86/36.83/-20.32/18.88/-39.12/-40.71/-0.97/
-23.64/0.08/

***cycle***
   Generation:  75  Best population fitness : 0.02428789 Mean of population:
0.03478878
    Best chromosome->10.70/-19.86/36.83/-20.32/18.88/-39.12/-40.71/-0.97/
-23.64/0.08/

***cycle***
   Generation:  76  Best population fitness : 0.02428789 Mean of population:
0.03472346
    Best chromosome->10.70/-19.86/36.83/-20.32/18.88/-39.12/-40.71/-0.97/
-23.64/0.08/

***cycle***
   Generation:  77  Best population fitness : 0.02301687 Mean of population:
0.03458719
    Best chromosome->27.32/-31.84/-3.00/-57.12/24.88/-16.14/-1.14/39.73/
-1.15/0.18/

***cycle***
   Generation:  78  Best population fitness : 0.02301687 Mean of population:
0.03451069
    Best chromosome->27.32/-31.84/-3.00/-57.12/24.88/-16.14/-1.14/39.73/-1.15/
0.18/

***cycle***
   Generation:  79  Best population fitness : 0.02301687 Mean of population:
0.03434609
    Best chromosome->27.32/-31.84/-3.00/-57.12/24.88/-16.14/-1.14/39.73/-1.15
/0.18/

***cycle***
```

    Generation: 80 Best population fitness : 0.02301687 Mean of population: 0.03418806
    Best chromosome->27.32/-31.84/-3.00/-57.12/24.88/-16.14/-1.14/39.73/-1.15/0.18/

    cycle
    Generation: 81 Best population fitness : 0.02301687 Mean of population: 0.03402732
    Best chromosome->27.32/-31.84/-3.00/-57.12/24.88/-16.14/-1.14/39.73/-1.15/0.18/

    cycle
    Generation: 82 Best population fitness : 0.02301687 Mean of population: 0.03390054
    Best chromosome->27.32/-31.84/-3.00/-57.12/24.88/-16.14/-1.14/39.73/-1.15/0.18/

    cycle
    Generation: 83 Best population fitness : 0.02301687 Mean of population: 0.03387266
    Best chromosome->27.32/-31.84/-3.00/-57.12/24.88/-16.14/-1.14/39.73/-1.15/0.18/

    cycle
    Generation: 84 Best population fitness : 0.02301687 Mean of population: 0.03376543
    Best chromosome->27.32/-31.84/-3.00/-57.12/24.88/-16.14/-1.14/39.73/-1.15/0.18/

    cycle
    Generation: 85 Best population fitness : 0.02301687 Mean of population: 0.03370612
    Best chromosome->27.32/-31.84/-3.00/-57.12/24.88/-16.14/-1.14/39.73/-1.15/0.18/

    cycle
    Generation: 86 Best population fitness : 0.02301687 Mean of population: 0.03360450
    Best chromosome->27.32/-31.84/-3.00/-57.12/24.88/-16.14/-1.14/39.73/-1.15/0.18/

    cycle
    Generation: 87 Best population fitness : 0.02301687 Mean of population: 0.03355560

Best chromosome->27.32/-31.84/-3.00/-57.12/24.88/-16.14/-1.14/39.73/-1.15/0.18/

cycle
Generation: 88 Best population fitness : 0.02301687 Mean of population: 0.03331063
Best chromosome->27.32/-31.84/-3.00/-57.12/24.88/-16.14/-1.14/39.73/-1.15/0.18/

cycle
Generation: 89 Best population fitness : 0.02301687 Mean of population: 0.03308697
Best chromosome->27.32/-31.84/-3.00/-57.12/24.88/-16.14/-1.14/39.73/-1.15/0.18/

cycle
Generation: 90 Best population fitness : 0.02301687 Mean of population: 0.03299376
Best chromosome->27.32/-31.84/-3.00/-57.12/24.88/-16.14/-1.14/39.73/-1.15/0.18/

cycle
Generation: 91 Best population fitness : 0.02301687 Mean of population: 0.03280551
Best chromosome->27.32/-31.84/-3.00/-57.12/24.88/-16.14/-1.14/39.73/-1.15/0.18/

cycle
Generation: 92 Best population fitness : 0.02152323 Mean of population: 0.03267332
Best chromosome->27.32/-41.85/4.38/-14.27/24.88/-14.09/-9.88/39.73/-1.15/0.18/

cycle
Generation: 93 Best population fitness : 0.02152323 Mean of population: 0.03265421
Best chromosome->27.32/-41.85/4.38/-14.27/24.88/-14.09/-9.88/39.73/-1.15/0.18/

cycle
Generation: 94 Best population fitness : 0.02152323 Mean of population: 0.03250503
Best chromosome->27.32/-41.85/4.38/-14.27/24.88/-14.09/-9.88/39.73/-1.15/0.18/

```
    ***cycle***
    Generation:  95  Best population fitness : 0.02152323 Mean of population:
0.03202693
     Best chromosome->27.32/-41.85/4.38/-14.27/24.88/-14.09/-9.88/39.73/-1.15/
0.18/

    ***cycle***
    Generation:  96  Best population fitness : 0.02002237 Mean of population:
0.03194529
     Best chromosome->0.29/-5.15/3.81/-11.03/17.01/-9.38/21.15/5.84/-0.76/0.17/

    ***cycle***
    Generation:  97  Best population fitness : 0.02002237 Mean of population:
0.03192516
     Best chromosome->0.29/-5.15/3.81/-11.03/17.01/-9.38/21.15/5.84/-0.76/0.17/

    ***cycle***
    Generation:  98  Best population fitness : 0.02002237 Mean of population:
0.03178475
     Best chromosome->0.29/-5.15/3.81/-11.03/17.01/-9.38/21.15/5.84/-0.76/0.17/

    ***cycle***
    Generation:  99  Best population fitness : 0.02002237 Mean of population:
0.03174589
     Best chromosome->0.29/-5.15/3.81/-11.03/17.01/-9.38/21.15/5.84/-0.76/0.17/

    ***cycle***
    Generation:  100  Best population fitness : 0.02002237 Mean of population:
0.03169513
     Best chromosome->0.29/-5.15/3.81/-11.03/17.01/-9.38/21.15/5.84/-0.76/0.17/
    Call:
    ANNGA.default(x = input, y = output, design = c(1, 3, 1), population = 100,
       mutation = 0.3, crossover = 0.7, maxGen = 100, error = 0.001)

    ************************************************************
    Mean Squared Error-------------------------------> 0.02002237
    R2-----------------------------------------------> 0.4627776
    Number of generation-----------------------------> 101
    Weight range at initialization-------------------> [ 25 , -25 ]
    Weight range resulted from the optimisation------> [ 21.14924 , -11.02609 ]
    ************************************************************
```

第 10 章

关联性规则

关联性规则最早是由 R.Agrawal 等人针对超市购物篮分析（Market Basket Analysis）问题提出的，其目的是发现超市的事务处理数据库中不同商品间的关联关系。关联性规则呈现了顾客购物的行为模式，其结果可以作为经营决策、市场预测和制定销售策略的参考依据。以尿布及啤酒的购物篮为例：

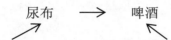

左半部（Left-Hand Side，LHS）　　右半部（Right-Hand Side，RHS）

此规则表示尿布和啤酒销售具有关联性。

假设总共有 N 笔交易，定义 supp（A）为购买项目 A 的支持度（Support），conf（A→B）为既会购买项目 A 又会购买项目 B 的置信度（Confidence）。支持度可用来判断规则的有效性，置信度用来判断在项目 A 的条件下发生项目 B 的可能性，其值越高，规则就越具有参考价值。一个强关联性规则通常支持度和置信度值都高，但反过来，支持度和置信度值都高却不一定代表此关联性规则所指的事件彼此间就一定存在着高相关性，同时，还需检查提升度（Lift）是否大于 1，提升度大于 1，表示项目 A 与项目 B 间有正向关系；提升度值等于 1，表示项目 A 与项目 B 间没有关系；提升度小于 1，表示项目 A 与项目 B 间为负向关系。支持度、置信度及提升度的计算方法如下：

```
supp(A)   = freq(A)/N
supp(A→B) = freq(A,B)/N
conf(A→B) = freq(A,B)/freq(A)
lift(A→B) = conf(A→B)/supp(B)
```

其中：
- freq(A)：购买项目 A 的数量。
- freq(A,B)：同时购买项目 A 及 B 的数量。

1994年，Agrawal & Srikant提出的Apriori算法，是目前常用的关联性规则。Apriori算法采用一种逐层搜索的迭代方法（Level-Wise Search），先找出满足最小支持度的频繁项目集（Frequent Itemsets），再以最小置信度为条件，计算频繁项目集所形成的关联性规则，当Apriori算法找出满足用户指定的最小支持度（Minimum Support）及最小置信度（Minimum Confidence）的关联性规则时，这个规则才成立。Apriori算法步骤如下：

步骤01 找出频繁项目集（Frequent Itemset）L_1。

重复步骤02和步骤03直到无新频繁项目集产生（k ≥ 1）。

步骤02 取得长度为k+1的候选项目集（Candidate Itemset）C_{k+1}。

- 组合（Join）：将L_k中的项目集两两组合为C'_{k+1}。
- 剪枝（Prune）：修剪子集合不属于L_k的候选项目集C'_{k+1}，得到长度为k+1的候选项目集C_{k+1}。

步骤03 找出长度为k+1的频繁项目集L_{k+1}。

- 计数（Count）：计算剪枝后候选项目集C_{k+1}的支持度。
- 删除（Delete）：删除支持度未达到最小支持度的候选项目集C_{k+1}，产生长度为k+1的频繁项目集L_{k+1}。

步骤04 由频繁项目集产生关联性规则。

2000年，Zaki提出了Eclat算法，这是一种深度优先算法，具体做法是将事务处理数据库中的项目（Item）作为键（Key），每个项目对应的交易编号（TID）作为值（Value）。Eclat算法如图10-1所示。

TID	Item
1	A, B
2	B, C
3	A, C
4	A, B, C

转换后

Item	TID
A	1, 3, 4
B	1, 2, 4
C	2, 3, 4

Item	Freq
A	3
B	3
C	3

计算长度为1的频繁项目集

Item	Freq
A, B	2
A, C	2
B, C	2

由长度为1的频繁项目集产生长度为2的频繁项目集

Item	Freq
A, B, C	1

由长度为2的频繁项目集产生长度为3的频繁项目集

图10-1 Eclat算法

10.1 产生关联性规则并排序

[程序范例 10-1]

首先引用 arules 包：

```
> library(arules)
```

引用 Adult 数据集：

```
> data("Adult")
```

设置支持度=0.5、置信度=0.9 及不显示运行过程的相关信息（verbose=F）：

```
# 调用apriori()函数
> rules <- apriori(Adult,parameter = list(supp = 0.5, conf = 0.9),control=list(verbose=F))
```

将 52 条关联性规则按置信度大小排序并显示出来：

```
# 关联性规则按置信度大小排序
> rules.sorted=sort(rules,by="confidence")
> inspect(rules.sorted)

    lhs                          rhs                      support
confidence  lift
  1 {hours-per-week=Full-time} => {capital-loss=None}     0.5606650
0.9582531  1.0052191
  2 {workclass=Private}        => {capital-loss=None}     0.6639982
0.9564974  1.0033773
  3 {workclass=Private,
     native-country=United-States} => {capital-loss=None} 0.5897179
0.9554818  1.0023119
  4 {capital-gain=None,
     hours-per-week=Full-time} => {capital-loss=None}     0.5191638
0.9550659  1.0018756
  5 {workclass=Private,
     race=White}               => {capital-loss=None}     0.5674829
0.9549683  1.0017732
  6 {workclass=Private,
     race=White,
     native-country=United-States} => {capital-loss=None} 0.5181401
0.9535418  1.0002768
  7 {}                         => {capital-loss=None}     0.9532779
0.9532779  1.0000000
  8 {workclass=Private,
```

```
       capital-gain=None}              => {capital-loss=None}       0.6111748
0.9529145 0.9996188
    9  {native-country=United-States} => {capital-loss=None}       0.8548380
0.9525461 0.9992323
    10 {workclass=Private,
        capital-gain=None,
        native-country=United-States} => {capital-loss=None}       0.5414807
0.9517075 0.9983526
    11 {race=White}                    => {capital-loss=None}       0.8136849
0.9516307 0.9982720
    12 {workclass=Private,
        race=White,
        capital-gain=None}             => {capital-loss=None}       0.5204742
0.9511000 0.9977153
    13 {race=White,
        native-country=United-States} => {capital-loss=None}       0.7490480
0.9504325 0.9970152
    14 {capital-gain=None}             => {capital-loss=None}       0.8706646
0.9490705 0.9955863
    15 {capital-gain=None,
        native-country=United-States} => {capital-loss=None}       0.7793702
0.9481891 0.9946618
    16 {race=White,
        capital-gain=None}             => {capital-loss=None}       0.7404283
0.9470983 0.9935175
    17 {sex=Male}                      => {capital-loss=None}       0.6331027
0.9470750 0.9934931
    18 {sex=Male,
        native-country=United-States} => {capital-loss=None}       0.5661316
0.9462068 0.9925823
    19 {race=White,
        sex=Male}                      => {capital-loss=None}       0.5564268
0.9457804 0.9921350
    20 {race=White,
        capital-gain=None,
        native-country=United-States} => {capital-loss=None}       0.6803980
0.9457029 0.9920537
    21 {race=White,
        sex=Male,
        native-country=United-States} => {capital-loss=None}       0.5113632
0.9442722 0.9905529
    22 {sex=Male,
        capital-gain=None}             => {capital-loss=None}       0.5696941
0.9415288 0.9876750
```

```
   23 {sex=Male,
      capital-gain=None,
      native-country=United-States} => {capital-loss=None}       0.5084149
0.9404636 0.9865576
   24 {hours-per-week=Full-time}      => {capital-gain=None}     0.5435895
0.9290688 1.0127342
   25 {capital-loss=None,
      hours-per-week=Full-time}       => {capital-gain=None}     0.5191638
0.9259787 1.0093657
   26 {workclass=Private}             => {capital-gain=None}     0.6413742
0.9239073 1.0071078
   27 {workclass=Private,
      native-country=United-States}   => {capital-gain=None}     0.5689570
0.9218444 1.0048592
   28 {race=White}                    => {native-country=United-States} 0.7881127
0.9217231 1.0270761
   29 {workclass=Private,
      race=White}                     => {capital-gain=None}     0.5472339
0.9208931 1.0038221
   30 {race=White,
      capital-loss=None}              => {native-country=United-States} 0.7490480
0.9205626 1.0257830
   31 {race=White,
      sex=Male}                       => {native-country=United-States} 0.5415421
0.9204803 1.0256912
   32 {workclass=Private,
      capital-loss=None}              => {capital-gain=None}     0.6111748
0.9204465 1.0033354
   33 {race=White,
      capital-gain=None}              => {native-country=United-States} 0.7194628
0.9202807 1.0254689
   34 {race=White,
      sex=Male,
      capital-loss=None}              => {native-country=United-States} 0.5113632
0.9190124 1.0240556
   35 {race=White,
      capital-gain=None,
      capital-loss=None}              => {native-country=United-States} 0.6803980
0.9189249 1.0239581
   36 {workclass=Private,
      capital-loss=None,
      native-country=United-States}   => {capital-gain=None}     0.5414807
0.9182030 1.0008898
```

```
    37  {}                                  => {capital-gain=None}              0.9173867
0.9173867 1.0000000
    38  {workclass=Private,
         race=White,
         capital-loss=None}                 => {capital-gain=None}              0.5204742
0.9171628 0.9997559
    39  {native-country=United-States}      => {capital-gain=None}              0.8219565
0.9159062 0.9983862
    40  {workclass=Private,
         race=White}                        => {native-country=United-States}   0.5433848
0.9144157 1.0189334
    41  {race=White}                        => {capital-gain=None}              0.7817862
0.9143240 0.9966616
    42  {capital-loss=None}                 => {capital-gain=None}              0.8706646
0.9133376 0.9955863
    43  {workclass=Private,
         race=White,
         capital-loss=None}                 => {native-country=United-States}   0.5181401
0.9130498 1.0174114
    44  {race=White,
         native-country=United-States}      => {capital-gain=None}              0.7194628
0.9128933 0.9951019
    45  {capital-loss=None,
         native-country=United-States}      => {capital-gain=None}              0.7793702
0.9117168 0.9938195
    46  {race=White,
         capital-loss=None}                 => {capital-gain=None}              0.7404283
0.9099693 0.9919147
    47  {race=White,
         capital-loss=None,
         native-country=United-States}      => {capital-gain=None}              0.6803980
0.9083504 0.9901500
    48  {sex=Male}                          => {capital-gain=None}              0.6050735
0.9051455 0.9866565
    49  {sex=Male,
         native-country=United-States}      => {race=White}                     0.5415421
0.9051090 1.0585540
    50  {sex=Male,
         native-country=United-States}      => {capital-gain=None}              0.5406003
0.9035349 0.9849008
    51  {sex=Male,
         capital-loss=None,
         native-country=United-States}      => {race=White}                     0.5113632
0.9032585 1.0563898
```

```
  52  {race=White,
       sex=Male}                  => {capital-gain=None}      0.5313050
0.9030799 0.9844048
```

设置支持度=0.5、置信度=0.9、不显示运行过程的相关信息及不包含 race=White 及 sex=Male 的项目集：

```
> is <- apriori(Adult, parameter = list(supp = 0.5, conf = 0.9),
+ appearance = list(none = c("race=White", "sex=Male")),
+ control=list(verbose=F))
```

确认没有 race=White 及 sex=Male 的项目集：

```
> itemFrequency(items(is))["race=White"]
race=White
         0
> itemFrequency(items(is))["sex=Male"]
sex=Male
       0
```

将 20 条关联性规则按置信度大小排序并显示出来：

```
# 关联性规则按置信度大小排序
> is.sorted=sort(is,by="confidence")
> inspect(is.sorted)

     lhs                            rhs                     support   confidence  lift
  1  {hours-per-week=Full-time}  => {capital-loss=None}   0.5606650  0.9582531
1.0052191
  2  {workclass=Private}         => {capital-loss=None}   0.6639982  0.9564974
1.0033773
  3  {workclass=Private,
      native-country=United-States} => {capital-loss=None} 0.5897179  0.9554818
1.0023119
  4  {capital-gain=None,
      hours-per-week=Full-time}  => {capital-loss=None}   0.5191638  0.9550659
1.0018756
  5  {}                          => {capital-loss=None}   0.9532779  0.9532779
1.0000000
  6  {workclass=Private,
      capital-gain=None}         => {capital-loss=None}   0.6111748  0.9529145
0.9996188
  7  {native-country=United-States} => {capital-loss=None} 0.8548380  0.9525461
0.9992323
  8  {workclass=Private,
      capital-gain=None,
```

```
       native-country=United-States} => {capital-loss=None} 0.5414807 0.9517075
0.9983526
    9  {capital-gain=None}            => {capital-loss=None} 0.8706646 0.9490705
0.9955863
    10 {capital-gain=None,
       native-country=United-States} => {capital-loss=None} 0.7793702 0.9481891
0.9946618
    11 {hours-per-week=Full-time}     => {capital-gain=None} 0.5435895 0.9290688
1.0127342
    12 {capital-loss=None,
       hours-per-week=Full-time}      => {capital-gain=None} 0.5191638 0.9259787
1.0093657
    13 {workclass=Private}            => {capital-gain=None} 0.6413742 0.9239073
1.0071078
    14 {workclass=Private,
       native-country=United-States} => {capital-gain=None} 0.5689570 0.9218444
1.0048592
    15 {workclass=Private,
       capital-loss=None}             => {capital-gain=None} 0.6111748 0.9204465
1.0033354
    16 {workclass=Private,
       capital-loss=None,
       native-country=United-States} => {capital-gain=None} 0.5414807 0.9182030
1.0008898
    17 {}                             => {capital-gain=None} 0.9173867 0.9173867
1.0000000
    18 {native-country=United-States} => {capital-gain=None} 0.8219565 0.9159062
0.9983862
    19 {capital-loss=None}            => {capital-gain=None} 0.8706646 0.9133376
0.9955863
    20 {capital-loss=None,
       native-country=United-States} => {capital-gain=None} 0.7793702 0.9117168
0.9938195
```

10.2 删除冗余规则

[程序范例 10-2]

首先引用 arules 包：

```
> library(arules)
```

使用 Titanic 数据集并转换为数据框对象 df：

```
> data("Titanic")
> str(Titanic)
 table [1:4, 1:2, 1:2, 1:2] 0 0 35 0 0 0 17 0 118 154 ...
 - attr(*, "dimnames")=List of 4
  ..$ Class   : chr [1:4] "1st" "2nd" "3rd" "Crew"
  ..$ Sex     : chr [1:2] "Male" "Female"
  ..$ Age     : chr [1:2] "Child" "Adult"
  ..$ Survived: chr [1:2] "No" "Yes"
> df <- as.data.frame(Titanic)
```

产生符合 apriori()函数要求的新数据框对象 titanic.new：

```
> titanic.new <- NULL
> for(i in 1:4) {
+    titanic.new <- cbind(titanic.new, rep(as.character(df[,i]),df$Freq))
+ }
>
> titanic.new <- as.data.frame(titanic.new)
> names(titanic.new) <- names(df)[1:4]
> str(titanic.new)
'data.frame':   2201 obs. of  4 variables:
 $ Class   : Factor w/ 4 levels "1st","2nd","3rd",..: 3 3 3 3 3 3 3 3 3 3 ...
 $ Sex     : Factor w/ 2 levels "Female","Male": 2 2 2 2 2 2 2 2 2 2 ...
 $ Age     : Factor w/ 2 levels "Adult","Child": 2 2 2 2 2 2 2 2 2 2 ...
 $ Survived: Factor w/ 2 levels "No","Yes": 1 1 1 1 1 1 1 1 1 1 ...
```

默认按最小支持度= 0.1、最小置信度=0.8 进行关联分析，获得 27 条规则：

```
> titanic_rules.all <- apriori(titanic.new)
Parameter specification:
 confidence minval smax arem  aval originalSupport support minlen maxlen target   ext
        0.8    0.1    1 none FALSE            TRUE     0.1      1     10  rules FALSE
Algorithmic control:
 filter tree heap memopt load sort verbose
    0.1 TRUE FALSE  TRUE    2    TRUE


apriori - find association rules with the apriori algorithm
version 4.21 (2004.05.09)        (c) 1996-2004   Christian Borgelt
set item appearances ...[0 item(s)] done [0.00s].
set transactions ...[10 item(s), 2201 transaction(s)] done [0.00s].
sorting and recoding items ... [9 item(s)] done [0.00s].
creating transaction tree ... done [0.00s].
checking subsets of size 1 2 3 4 done [0.00s].
writing ... [27 rule(s)] done [0.00s].
creating S4 object  ... done [0.00s].
```

```
> inspect(titanic_rules.all)
   lhs                rhs              support  confidence lift
1  {}              => {Age=Adult}      0.9504771 0.9504771 1.0000000
2  {Class=2nd}     => {Age=Adult}      0.1185825 0.9157895 0.9635051
3  {Class=1st}     => {Age=Adult}      0.1449341 0.9815385 1.0326798
4  {Sex=Female}    => {Age=Adult}      0.1930940 0.9042553 0.9513700
5  {Class=3rd}     => {Age=Adult}      0.2848705 0.8881020 0.9343750
6  {Survived=Yes}  => {Age=Adult}      0.2971377 0.9198312 0.9677574
7  {Class=Crew}    => {Sex=Male}       0.3916402 0.9740113 1.2384742
8  {Class=Crew}    => {Age=Adult}      0.4020900 1.0000000 1.0521033
9  {Survived=No}   => {Sex=Male}       0.6197183 0.9154362 1.1639949
10 {Survived=No}   => {Age=Adult}      0.6533394 0.9651007 1.0153856
11 {Sex=Male}      => {Age=Adult}      0.7573830 0.9630272 1.0132040
12 {Sex=Female,
    Survived=Yes} => {Age=Adult}      0.1435711 0.9186047 0.9664669
13 {Class=3rd,
    Sex=Male}     => {Survived=No}    0.1917310 0.8274510 1.2222950
14 {Class=3rd,
    Survived=No}  => {Age=Adult}      0.2162653 0.9015152 0.9484870
15 {Class=3rd,
    Sex=Male}     => {Age=Adult}      0.2099046 0.9058824 0.9530818
16 {Sex=Male,
    Survived=Yes} => {Age=Adult}      0.1535666 0.9209809 0.9689670
17 {Class=Crew,
    Survived=No}  => {Sex=Male}       0.3044071 0.9955423 1.2658514
18 {Class=Crew,
    Survived=No}  => {Age=Adult}      0.3057701 1.0000000 1.0521033
19 {Class=Crew,
    Sex=Male}     => {Age=Adult}      0.3916402 1.0000000 1.0521033
20 {Class=Crew,
    Age=Adult}    => {Sex=Male}       0.3916402 0.9740113 1.2384742
21 {Sex=Male,
    Survived=No}  => {Age=Adult}      0.6038164 0.9743402 1.0251065
22 {Age=Adult,
    Survived=No}  => {Sex=Male}       0.6038164 0.9242003 1.1751385
23 {Class=3rd,
    Sex=Male,
    Survived=No}  => {Age=Adult}      0.1758292 0.9170616 0.9648435
24 {Class=3rd,
    Age=Adult,
    Survived=No}  => {Sex=Male}       0.1758292 0.8130252 1.0337773
25 {Class=3rd,
    Sex=Male,
    Age=Adult}    => {Survived=No}    0.1758292 0.8376623 1.2373791
```

```
26 {Class=Crew,
    Sex=Male,
    Survived=No}   => {Age=Adult}     0.3044071  1.0000000  1.0521033
27 {Class=Crew,
    Age=Adult,
    Survived=No}   => {Sex=Male}      0.3044071  0.9955423  1.2658514
```

若设置按最小支持度=0.005、最小置信度=0.8 进行关联分析，可获得 12 条关联性规则并按照提升度大小来排序：

```
> rules <- apriori(titanic.new, control = list(verbose=F),
+ parameter = list(minlen=2, supp=0.005, conf=0.8),
+ appearance = list(rhs=c("Survived=No", "Survived=Yes"),
+ default="lhs"))
> quality(rules) <- round(quality(rules), digits=3)

# 关联性规则按照提升度大小排序
> rules.sorted <- sort(rules, by="lift")
> inspect(rules.sorted)
   lhs                rhs              support confidence lift
1  {Class=2nd,
    Age=Child}     => {Survived=Yes}   0.011    1.000   3.096
2  {Class=2nd,
    Sex=Female,
    Age=Child}     => {Survived=Yes}   0.006    1.000   3.096
3  {Class=1st,
    Sex=Female}    => {Survived=Yes}   0.064    0.972   3.010
4  {Class=1st,
    Sex=Female,
    Age=Adult}     => {Survived=Yes}   0.064    0.972   3.010
5  {Class=2nd,
    Sex=Female}    => {Survived=Yes}   0.042    0.877   2.716
6  {Class=Crew,
    Sex=Female}    => {Survived=Yes}   0.009    0.870   2.692
7  {Class=Crew,
    Sex=Female,
    Age=Adult}     => {Survived=Yes}   0.009    0.870   2.692
8  {Class=2nd,
    Sex=Female,
    Age=Adult}     => {Survived=Yes}   0.036    0.860   2.663
9  {Class=2nd,
    Sex=Male,
    Age=Adult}     => {Survived=No}    0.070    0.917   1.354
10 {Class=2nd,
    Sex=Male}      => {Survived=No}    0.070    0.860   1.271
```

```
11 {Class=3rd,
   Sex=Male,
   Age=Adult}     => {Survived=No}    0.176      0.838 1.237
12 {Class=3rd,
   Sex=Male}      => {Survived=No}    0.192      0.827 1.222
```

在产生关联性规则的结果中，某些规则比其他规则提供的信息少或没有额外信息，则称该条规则为冗余（Redundancy）规则。一般而言，冗余规则的提升度与其相关的关联性规则的提升度相同或者较低。由以下结果可知有4条冗余规则。

```
# 去除冗余规则
> subset.matrix <- is.subset(rules.sorted, rules.sorted)
> subset.matrix[lower.tri(subset.matrix, diag=T)] <- NA
> redundant <- colSums(subset.matrix, na.rm=T) >= 1
> which(redundant)
[1] 2 4 7 8
```

去除冗余规则后，得到12条关联性规则：

```
# 去除冗余规则
> rules.pruned <- rules.sorted[!redundant]
> inspect(rules.pruned)
  lhs              rhs              support confidence lift
1 {Class=2nd,
   Age=Child}    => {Survived=Yes}   0.011      1.000 3.096
2 {Class=1st,
   Sex=Female}   => {Survived=Yes}   0.064      0.972 3.010
3 {Class=2nd,
   Sex=Female}   => {Survived=Yes}   0.042      0.877 2.716
4 {Class=Crew,
   Sex=Female}   => {Survived=Yes}   0.009      0.870 2.692
5 {Class=2nd,
   Sex=Male,
   Age=Adult}    => {Survived=No}    0.070      0.917 1.354
6 {Class=2nd,
   Sex=Male}     => {Survived=No}    0.070      0.860 1.271
7 {Class=3rd,
   Sex=Male,
   Age=Adult}    => {Survived=No}    0.176      0.838 1.237
8 {Class=3rd,
   Sex=Male}     => {Survived=No}    0.192      0.827 1.222
```

我们可针对lhs进行细部调整，借以进一步分析关联性规则。本范例中针对lhs的Class=1st、Class=2nd、Class=3rd及Age=Child、Age=Adult进一步分析并产生6条关联性规则。

```
> rules <- apriori(titanic.new,
+ parameter = list(minlen=3, supp=0.002, conf=0.2),
```

```
+ appearance = list(rhs=c("Survived=Yes"),
+ lhs=c("Class=1st", "Class=2nd", "Class=3rd",
+ "Age=Child", "Age=Adult"),
+ default="none"),
+ control = list(verbose=F))

# 关联性规则按照置信度大小排序
> rules.sorted <- sort(rules, by="confidence")
> inspect(rules.sorted)
  lhs              rhs              support    confidence lift
1 {Class=2nd,
   Age=Child}  => {Survived=Yes}    0.010904134 1.0000000  3.0956399
2 {Class=1st,
   Age=Child}  => {Survived=Yes}    0.002726034 1.0000000  3.0956399
3 {Class=1st,
   Age=Adult}  => {Survived=Yes}    0.089504771 0.6175549  1.9117275
4 {Class=2nd,
   Age=Adult}  => {Survived=Yes}    0.042707860 0.3601533  1.1149048
5 {Class=3rd,
   Age=Child}  => {Survived=Yes}    0.012267151 0.3417722  1.0580035
6 {Class=3rd,
   Age=Adult}  => {Survived=Yes}    0.068605179 0.2408293  0.7455209
```

我们可再引用 arulesViz 包及调用 plot()函数来产生关联性规则散布图和 Two-Key 图。

```
> library(arulesViz)
> plot(titanic_rules.all)
> plot(titanic_rules.all, shading="order", control=list(main =
+     "Two-key plot",col=rainbow(5)))
```

结果如图 10-2 和图 10-3 所示。

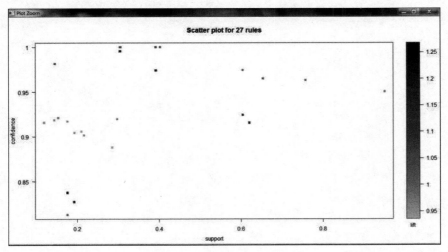

图 10-2　关联性规则散布图

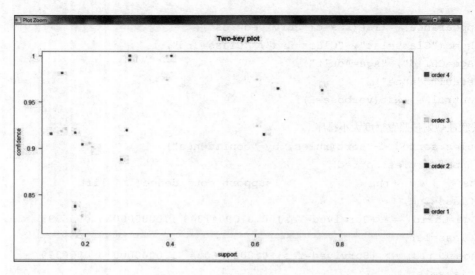

图 10-3　Two-Key 图

10.3　习题

调用 apriori()函数分析 insurance.csv。

第 11 章

文 本 挖 掘

文本挖掘（Text Mining）不同于数据挖掘的地方在于，它是要挖掘没有特定结构的纯文本，而这些文本的内容数据中也可能蕴藏着有用的信息。

11.1 使用混合分词并创建词频表

[程序范例 11-1]

首先引用 gutenbergr、jiebaR、dplyr 及 wordcloud 包：

```
> library(gutenbergr)
> library(jiebaR)
> library(dplyr)
> library(wordcloud)
```

调用 jiebaR 包中 worker() 函数默认的混合分词类型：

```
> mixSeg <- worker()
```

下载 PG 网站（https://www.gutenberg.org/）中编号为 64924-0 的电子书并赋值给 luxun 对象：

```
> luxun <- gutenberg_download(64924-0)
```

调用 segment() 函数对 luxun 对象的 text 变量分词并赋值给 luxun.seg 对象：

```
> str(luxun)
tibble[,2] [4,684 x 2] (S3: tbl_df/tbl/data.frame)
 $ gutenberg_id: int [1:4684] 64924 64924 64924 64924 64924 64924 64924 64924 64924 64924 ...
```

```
    $ text       : chr [1:4684] "[Illustration:" "" " "THEY CALLED ACROSS MERRILY
 TO EACH OTHER"" "]" ...
```

```
> luxun.seg <- segment(luxun$text, mixSeg)
```

调用 head()函数显示前 6 笔数据：

```
> luxun_head <- head(luxun.seg)
> luxun_head
```

调用 freq()函数建立词频表并加上列名称：

```
> luxun.freq <- freq(luxun.seg)
> colnames(luxun.freq) <- c("word","freq")
```

调用 arrange()函数制作一个排序过（从高至低）的频率表，显示前 6 笔数据：

```
> freq_df <- arrange(luxun.freq, desc(freq))
> head(freq_df)
  word freq
1  the 1648
2   to  757
3  and  757
4   of  685
5    a  614
6   in  494
```

11.2 使用 tag 分词并创建词云

使用 tag 重新分词、标注词性并赋值给 luxun.pos 对象：

```
> pos.tagger <- worker("tag")
> luxun.pos <- segment(luxun$text, pos.tagger)
```

调用 name()函数取得词性后与 luxun.pos 对象建立数据框，加上列名称：

```
> tmp_df <- data.frame(luxun.pos, names(luxun.pos))
> colnames(tmp_df)<-c("Word","POS")
```

启用管道（Pipe）符号 %>% 功能，将tmp_df传送给groupBy()函数来群组化Word及POS行数据，再使用%>%传送至summarize()及n()函数来计算群组的总笔数，最后将结果赋值给Word_POS_Freq。

```
> tmp_df %>%
+   group_by(Word,POS) %>%
+   summarise(Frequency=n()) -> Word_POS_Freq
```

调用 arrange()函数制作一个排序过（从高至低）的词频表，显示前 6 笔数据：

```
> Word_POS_Freq <- arrange(Word_POS_Freq, desc(Frequency))
> head(Word_POS_Freq)
# A tibble: 6 x 3
# Groups:   Word [6]
  Word  POS   Frequency
  <chr> <chr>     <int>
1 13    m             1
2 15    m             1
3 170   m             1
4 1906  m             1
5 1906. m             1
6 2     x             1
```

调用 subset()函数取出 POS 为代名词（标记为 r）的前 200 个词，计算出各个词的频率后再调用 wordcloud()函数把结果显示出来。

```
> pos_eng <- head(subset(Word_POS_Freq,Word_POS_Freq$POS == "eng"),100)
> pos_eng
# A tibble: 100 x 3
# Groups:   Word [100]
   Word     POS   Frequency
   <chr>    <chr>     <int>
 1 abandon  eng           1
 2 able     eng           2
 3 about    eng          69
 4 About    eng           2
 5 above    eng           6
 6 abroad   eng           1
 7 absence  eng           4
 8 absent   eng           2
 9 absolute eng           1
10 absorbed eng           2
# ... with 90 more rows

> pos_eng_barplot <- ggplot(aes(x = Word, y = Frequency), data =
+ pos_eng)+ geom_bar(stat="identity")
> pos_eng_barplot
```

频率如图 11-1 所示。

调用 wordcloud()函数来显示词云信息，如图 11-2 所示。

```
> wordcloud(pos_r$Word,pos_r$Frequency)
```

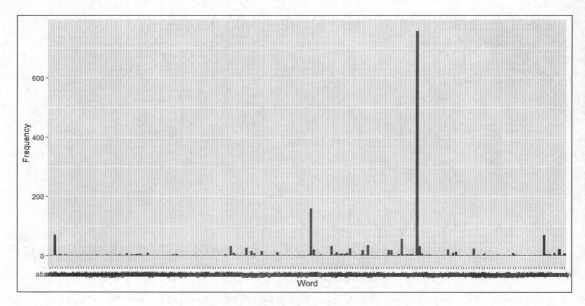

图 11-1 显示词频

图 11-2 显示标记为 eng 的词云

11.3 习题

到 https://www.gutenberg.org/files/64924/64924-0.txt 下载文本并分析。

第 12 章

推 荐 系 统

推荐系统是用于信息过滤的一种应用。推荐系统能够将可能购买（喜欢）的信息或物品推荐给用户，一方面帮助用户发现对自己有价值的信息或物品，另一方面让信息或物品能够推荐给其他可能购买（喜欢）的用户。

推荐系统中常根据用户与其他已经购买（喜欢）物品的用户之间的关联性来进行协同过滤推荐。对于协同过滤推荐，R 语言的 recommenderlab 包提供：（1）基于用户的协同过滤（User-Based Collaborative Filtering，UBCF），主要是当一个目标用户需要推荐时，可以先找到和他有相似购买（喜欢）的其他用户，然后把那些其他用户感兴趣而目标用户没有购买（喜欢）过的物品推荐给他；（2）基于物品的协同过滤（Item-Based Collaborative Filtering，IBCF），是推荐之前已经购买（喜欢）过的相似物品给目标用户，若物品 A 和物品 B 具有很大的相似度，则购买（喜欢）物品 A 的用户也可能会购买（喜欢）物品 B。

12.1 Jester5k 数据集

[程序范例 12-1]

首先引用 recommenderlab 包：

```
> library(recommenderlab)
```

使用 Jester5k 数据集，该数据集包含来自 1999 年 4 月至 2003 年 5 月收集的 Jester 在线笑话推荐人系统的 5000 名用户样本、100 个笑话，评分从-10 到 10 分，所有选定的用户都评了 36 个以上的笑话。

```
> data(Jester5k)
> Jester5k
```

```
5000 x 100 rating matrix of class 'realRatingMatrix' with 362106 ratings.
```

先调用 getRatings()函数取得评分值，再调用 hist()函数显示直方图，如图 12-1 所示。

```
> hist(getRatings(Jester5k))
```

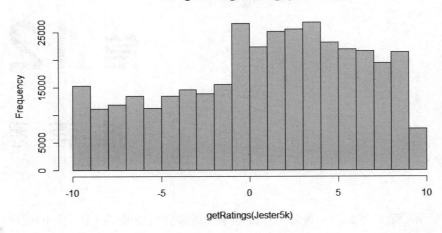

图 12-1　显示评分值的直方图

先调用 rowCounts()函数取得用户评了多少个笑话，再调用 summary()函数取得数据及分布信息。

```
> summary(rowCounts(Jester5k))
   Min. 1st Qu.  Median    Mean 3rd Qu.    Max.
  36.00   53.00   72.00   72.42  100.00  100.00
```

调用 evaluationScheme(data=Jester5k[1:1000], method="split", train=0.8, given=10, goodRating = 5)函数建立数据、抽样等计划，其中 data 是数据名，method= "cross-validation"是指使用 K-fold 分割数据；method="split"是指把数据分割为训练集和测试集；train=0.8 表示 80%数据是训练集；20%是测试集；given=10 是测试集中 10 个项目用于推荐算法，剩余的项目用于计算误差；goodRating=5 用于确定 5 分以上是好的评分。

```
> scheme= evaluationScheme(Jester5k[1:1000], method="split", train=0.8,
given=10, goodRating=5)
> scheme
Evaluation scheme with 10 items given
Method: 'split' with 1 run(s).
Training set proportion: 0.800
Good ratings: >=5.000000
Data set: 1000 x 100 rating matrix of class 'realRatingMatrix' with 72358 ratings.
```

调用 evaluate(scheme, method = "UBCF", type="topNList", n=c(1, 3, 5))函数来评估推荐结果，其中 method 是使用的推荐算法，type="topNList"是使用 TopN 推荐，n=c(1, 3, 5)表示推荐 Top1、Top3 及 Top5，type="ratings"是使用评分推荐。

```
> recN <- evaluate(scheme, method = "UBCF", type="topNList", n=c(1,3,5))
> recN
Evaluation results for 1 folds/samples using method 'UBCF'.
> recNavg=avg(recN)
> recNavg
      TP    FP    FN     TN  precision    recall        TPR        FPR
1  0.150 0.850 14.515 74.485     0.150 0.01358334 0.01358334 0.01147438
3  0.555 2.445 14.110 72.890     0.185 0.04308400 0.04308400 0.03267893
5  1.070 3.930 13.595 71.405     0.214 0.08396216 0.08396216 0.05238502
> plot(recN)
```

结果如图 12-2 所示。

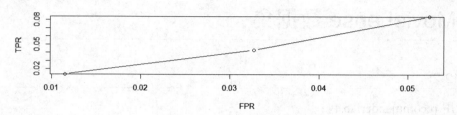

图 12-2　显示 topNList 评估结果

```
> recR <- evaluate(scheme, method = "UBCF", type="ratings")
> recR
> plot(recR)
```

结果如图 12-3 所示。

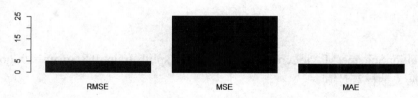

图 12-3　显示 ratings 评估的误差结果

可以调用 evaluate() 函数来进行 Z-score 标准化、选择 Cosine 相似度，同时评估 UBCF、IBCF 算法，然后调用 plot() 函数设置 ROC 或 prec/rec 参数来显示 ROC、precision 及 recall 值，调用 plot() 函数设置 annotate=1:3 来显示注释值。

```
> algorithms <- list(
+   "UBCF" = list(name="UBCF", param=list(normalize = "Z-score",
+             method="Cosine")),
+   "IBCF" = list(name="IBCF", param=list(normalize = "Z-score",
+             method="Cosine")))
> reclist <- evaluate(scheme, method = algorithms, type =
+             "topNList",n=c(1,3,5))
> reclist
> plot(reclist, "prec/rec", annotate=1:3,legend="bottomright")
```

结果如图 12-4 所示。

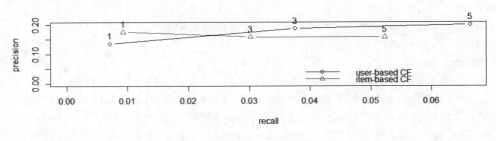

图 12-4　显示 precision 及 recall 值

12.2　MovieLense 数据集

[程序范例 12-2]

首先引用 recommenderlab 包：

```
> library(recommenderlab)
```

MovieLense 数据集是在 1997 年 9 月 19 日至 1998 年 4 月 22 日的 7 个月间通过 MovieLens 网站（movielens.umn.edu）收集的。该数据集包含 943 个用户对来自 1664 部电影的评分，评分从 1 分到 5 分。

```
> data(MovieLense)
> dim(MovieLense)
[1]  943 1664
```

调用 evaluationScheme(MovieLense[1:50], method="split", train=0.8, given=10, goodRating=5) 函数来建立数据、抽样等计划。

```
> scheme <- evaluationScheme(MovieLense[1:50], method = "split",
+                            train = 0.8, given = 10, goodRating = 5)
```

调用 Recommender(getData(scheme,"train"), method= "IBCF") 函数建立模型，其中 getData(scheme, "train") 表示训练集，method= "IBCF" 表示使用基于物品的协同过滤。

```
> model.ibcf <- Recommender(getData(scheme, "train"), method= "IBCF")
```

对已知部分的测试数据集（每个用户对 given=10 个物品评分）用 IBCF 预测评分。

```
> predict.ibcf <- predict(model.ibcf, getData(scheme, "known"), type
+                         = "ratings")
```

对未知部分的测试数据集调用 calcPredictionAccuracy() 函数计算误差。

```
> predict.error <- calcPredictionAccuracy(predict.ibcf,getData(scheme,
+                         "unknown"))
```

```
> predict.error
    RMSE      MSE      MAE
1.484331 2.203237 1.075540
```

使用 MovieLense 数据集中总评分大于 50 的前 50 个用户的数据来建立训练集。

```
> MovieLense50 <- MovieLense[rowCounts(MovieLense) >50,]
> train <- MovieLense50[1:50]
```

使用 IBCF 建立模型。

```
> model.ibcf <- Recommender(train, method = "IBCF")
> model.ibcf
Recommender of type 'IBCF' for 'realRatingMatrix'
learned using 50 users.
```

对 101 位和 102 位用户的数据进行预测评分并转成矩阵。

```
> predict.ibcf <- predict(model.ibcf, MovieLense100[101:102], type=
+                "ratings")
> as(predict.ibcf, "matrix")[,1:3]
    Toy Story (1995) GoldenEye (1995) Four Rooms (1995)
177               NA         3.666667                 3
178               NA               NA                 4
```

对 101 位和 102 位用户的数据预测 Top3 并转成列表。

```
> pre <- predict(rec, MovieLense100[101:102], n = 3)
> pre
Recommendations as 'topNList' with n = 3 for 2 users.
> as(pre, "list")
$`177`
[1] "Shanghai Triad (Yao a yao yao dao waipo qiao) (1995)"
[2] "Citizen Ruth (1996)"
[3] "unknown"

$`178`
[1] "Weekend at Bernie's (1989)" "Glengarry Glen Ross (1992)" "Audrey Rose (1977)"
```

第 13 章

可视化数据分析

rattle（The R Analytic Tool To Learn Easily）包是一个可视化的数据分析工具，允许我们从 CSV 文件、ARFF 或使用 ODBC（Open Database Connectivity，开放数据库互连）从 R Dataset、RData File、Library 或 Corpus 等来导入、探索（Explore）及测试（Test）、转换（Transform）数据，并建立（Build）和评估（Evaluate）模型，并可导出模型来供用户自行修改和运用。

首先引用 rattle 包并运行 rattle()函数：

```
> library('rattle')
> rattle()
```

rattle 运行窗口如图 13-1 所示（注意此窗口与 RStudio 或 R 窗口不同，用户须自行切换窗口）。

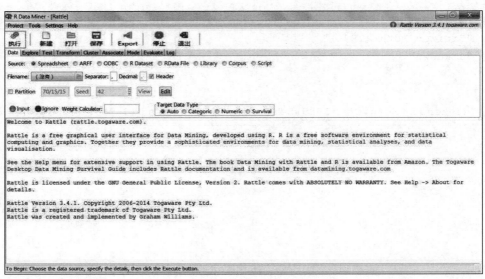

图 13-1　rattle 运行窗口

13.1 导入数据

我们可使用 rattle 来导入多种数据格式,数据格式包含 Spreadsheet、ARFF、ODBC、R Dataset、RData File、Library 及 Corpus。我们可使用 Spreadsheet 导入 CSV 及文本文件。若想导入 iris.csv 文件,其步骤为:(1)选择 Data,(2)选择 Source:Spreadsheet,(3)下拉 Filename 菜单,(4)选择 CSV Files 并选取 iris.csv 文件,(5)单击运行。操作顺序如图 13-2 所示。注意,可先按照生成 iris.csv 后再复制至 C:\下,并在 R 窗口运行以下程序,设置路径后再运行上述步骤。

```
> setwd("c:/")
> getwd()
[1] "c:/"
```

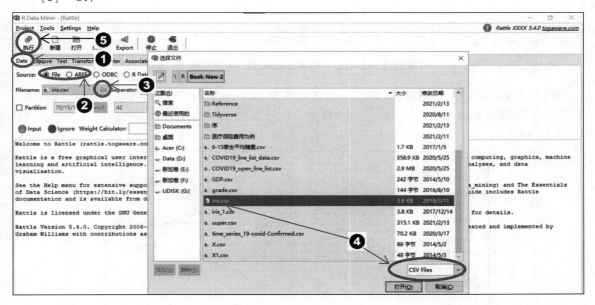

图 13-2 在 rattle 中导入 iris.csv

我们在导入 CSV 文件时可设置分隔符(Separator)及 Decimal 符号,也可决定是否导入 Header。图 13-3 所示为导入 iris.csv 时设置 Separator 为",",Decimal 为"."及要导入 Header 的操作界面。

ARFF 文件是 Weka 默认的数据格式(.arff),采用文本文件的格式。若要导入 iris.arff 文件,可参考导入 iris.csv 文件的方式进行(可至网址 https://github.com/renatopp/arff-datasets/blob/master/classification/iris.arff 下载 iris.arff 文件或使用本书提供的 iris.arff)。

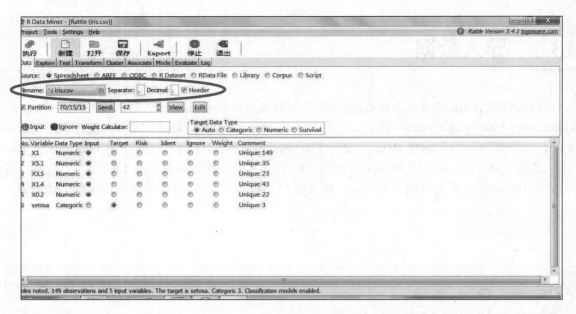

图 13-3　在 rattle 中导入 iris.csv 并设置相关参数

在使用 ODBC 导入数据库数据时，依次选择 Windows 下的"控制面板"→"系统管理工具"→"数据源（ODBC）"，再单击"新增"按钮来建立新的数据源，建立新数据源的界面如图 13-4 所示。

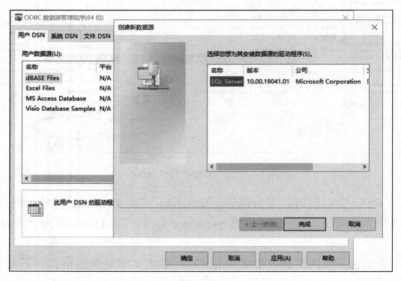

图 13-4　设置 ODBC 建立新数据源

在使用 ODBC 之前需先至 R 窗口中引用 RODBC 包：

```
> library(RODBC)
```

建立好数据源及引用 RODBC 包后，则可在 rattle 窗口中输入 DSN 名称（如 DSN：mitopac），然后按 Enter 键来选择数据库的数据表（Table），最后单击图 13-5 中的"执行"按钮就可以显示出数据。

第 13 章 可视化数据分析

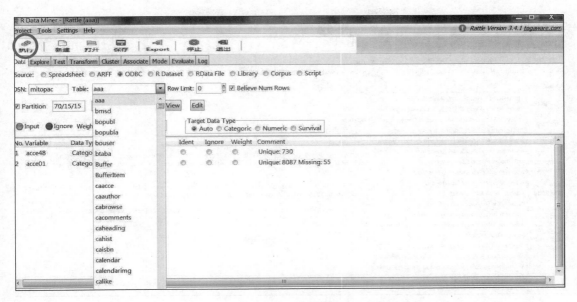

图 13-5 使用 ODBC 导入数据库中的数据

我们可以使用 R Dataset 来导入 SPSS、SAS 及 DBF 等数据，使用 Corpus 导入 Corpora 数据库中的数据。可使用 RData File 来导入 R 的 RData 格式文件，若要导入 iris.RData 文件，则可（1）选择 Data，（2）选择 Source：RData File，（3）下拉 Filename 菜单，选择 iris.RData 并设置 Data Name：iris，（4）单击"执行"按钮。操作界面如图 13-6 所示。

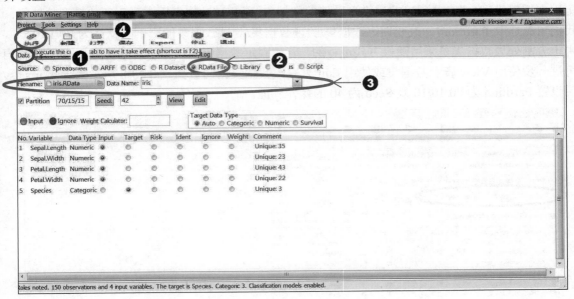

图 13-6 在 rattle 中导入 iris.RData

我们可使用 Library 来导入 R 程序中各个包的数据集，如要使用 arules 的 Adult 数据集，则可先下拉 Data Name 菜单并选择 AdultUCI：arules：Adult Data Set 选项，最后单击"执行"按钮。操作顺序如图 13-7 所示。

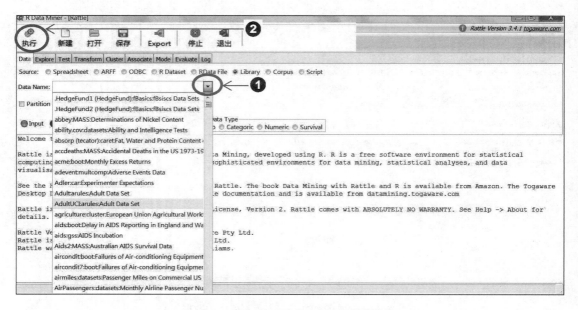

图 13-7　在 rattle 中导入 Adult 数据集

13.1.1　处理数据集

导入数据集后，rattle 可再分割处理数据集为训练数据集（Training Dataset）、验证数据集（Validation Dataset）及测试数据集（Testing Dataset）。训练数据集用来建立模型，验证数据集用来调整模型参数，是用以改进模型性能的数据集，而测试数据集用来真正评估模型的性能。我们可先选取 Partition 并改变训练数据集/验证数据集/测试数据集的比例（默认为 70/15/15），由于分割处理数据集是采用随机方式进行的，若想固定数据集的内容，则可设置 Seed 值；我们可进一步使用 View 查看数据集或使用 Edit 来修改数据集内的数据。图 13-8 所示为导入 iris.csv 后设置 Partition 为 80/10/10 及 Seed 为 40 的操作界面。

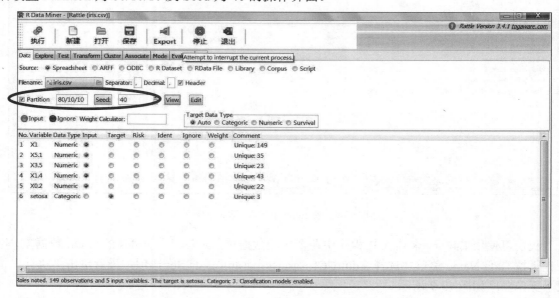

图 13-8　导入 iris.csv 后设置 Partition 及 Seed 的操作界面

13.1.2 设置变量

开始数据分析前，必须了解 rattle 提供的不同类型的变量（属性）的意义，rattle 变量可以是模型的输入（Input）变量，或者构建模型的目标（Target）变量（输出变量）。变量还可以细分为风险（Risk）变量，即不建议使用此变量用于构建模型；忽略（Ignore）变量，构建模型时先暂时忽略不使用的变量；识别（Ident）变量，具有唯一性的标识符（例如日期及身份证）；权重（Weight）变量，可设置不同权值的变量，我们可使用 Weight Calculator 设置变量权重，例如 abs(X1)/max(X1)*10+1 表示设置变量 X1 的权重。大多数变量的默认作用都是输入变量（自变量、独立变量、解释变量）。目标变量用于构建模型的输出值（因变量）。对于变量的数据类型（Data Type），可以是数值（Numeric）类型或类别（Categoric）类型，rattle 在导入数据时会默认将数值类型的变量设为目标变量，而类别类型的目标变量不适用于构建回归分析。我们可通过单选按钮（Radio Button）来选择符合需要的变量及数据类型，例如设置 iris.csv 变量时，X1 变量的权重为 abs(X1)/max(X1)*10+1，X5.1、X3.5、X1.4 及 X0.2 为数值类型的输入变量，setosa 为类别类型的目标变量。图 13-9 所示为设置 iris.csv 变量的界面。

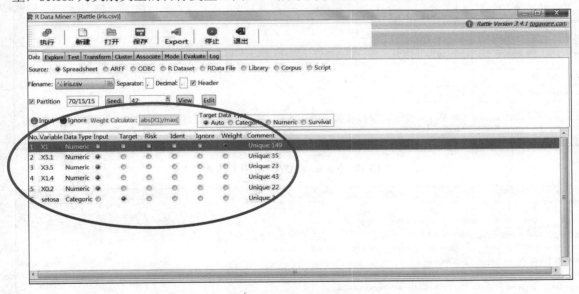

图 13-9 设置 iris.csv 变量的操作界面

13.2 探索及测试数据

导入数据后，rattle 提供摘要（Summary）分析、分布（Distributions）分析、相关（Correlation）分析及主成分（Principle Component）分析来探索了解数据。数据的摘要分析可探索了解数据集中每个变量的信息。对于数值类型的变量的摘要分析，可以包括最小值、最大值、中位数（Median）、平均值（Mean）和第一个及第三个四分位数（Quartile）；对于类别类型变量的摘要分析则提供频率分布（Frequency Distributions）。以 iris.csv 为例，运行摘要分析的步骤为：（1）选择 Explore，（2）选择 Summary，（3）单击"执行"按钮。图 13-10 所示为 iris.csv 摘要分析的执行界面。分布分析提供各种图形来了解数据的分布信息,其图形包含盒形图(Box Plot)及直方图(Histogram)等。图 13-11 所示为设置 iris.csv 中的 X1 变量运行盒形图进行分布分析,rattle 的图形会显示在 RStudio 或 R 窗口上,如图 13-12 所示。相关分析提供数值类型变量的相关性计算,并使用圆圈(Circle)和颜色来显示相关的强度,如图 13-13 和图 13-14 所示。主成分分析表示经由线性组合而得的主成分能保证原来变量最多的信息,即主成分有最大的方差且会显示出最大的个别差异,如图 13-15 和图 13-16 所示。

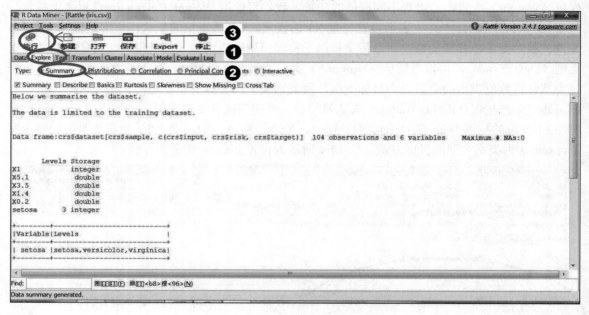

图 13-10　iris.csv 摘要分析的执行界面

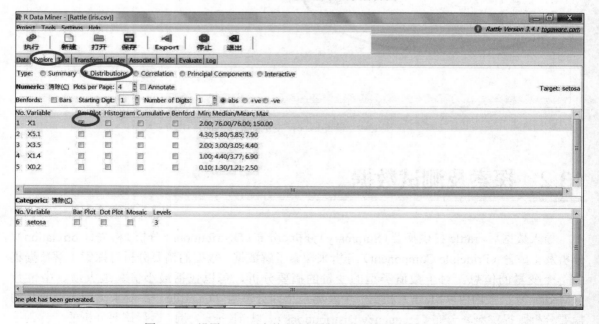

图 13-11　设置 iris.csv 中的 X1 变量执行盒形图进行分布分析

第 13 章 可视化数据分析

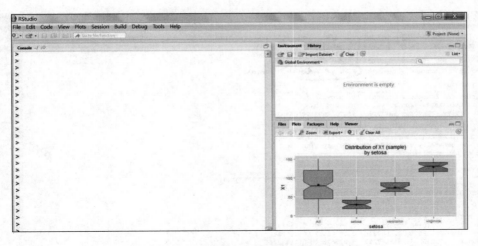

图 13-12　iris.csv 中 X1 变量的盒形图分布分析

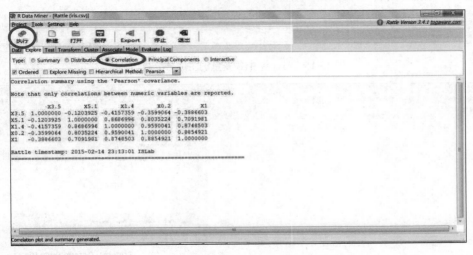

图 13-13　iris.csv 中各变量的相关分析值

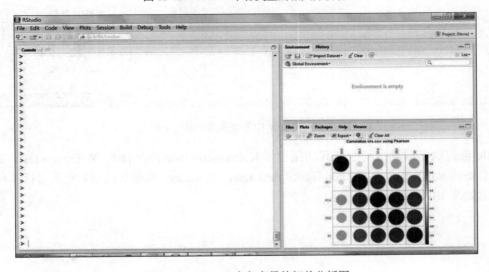

图 13-14　iris.csv 中各变量的相关分析图

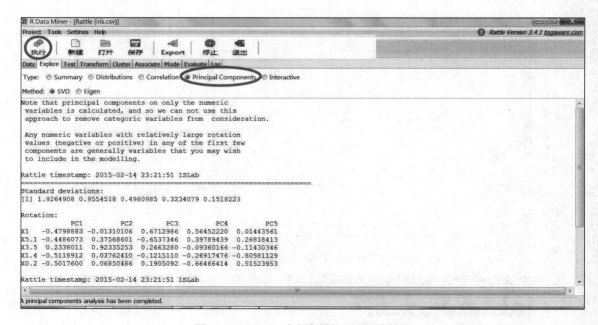

图 13-15　iris.csv 中各变量的主成分分析值

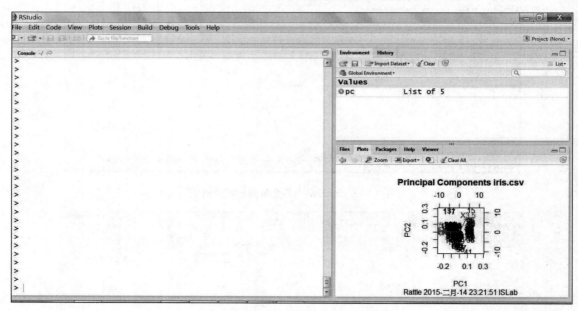

图 13-16　iris.csv 中各变量的主成分分析图

rattle 也提供多项测试功能，其功能包含 Kolmogorov-Smirnov test、Wilcoxon test、T-test、F-test、Correlation test 及 Wilcoxon signed rank test，以 iris.csv 数据集的 X1 变量运行 T-test，其执行界面如图 13-17 所示。

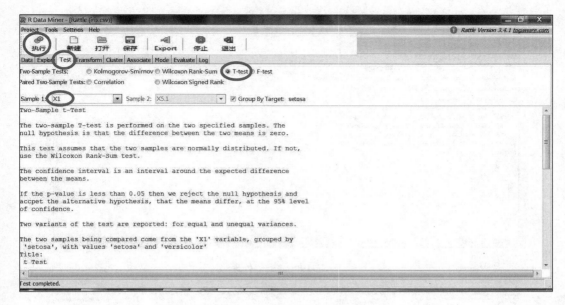

图 13-17 执行 iris.csv 数据集 X1 变量的 T-test 界面

13.3 转换数据

数据集中的数据可能存在缺失值，若数据中存在缺失值，则会严重影响数据分析的结果，甚至会影响建立模型的正确性。有些数据不能直接使用，必须进行转换，以确保模型的质量。rattle 中为数据的转换提供了重新缩放（Rescale）、缺失值处理（Impute）、转换数据类型（Recode）及清理数据（Cleanup）。转换数据的功能如图 13-18～图 13-21 所示。

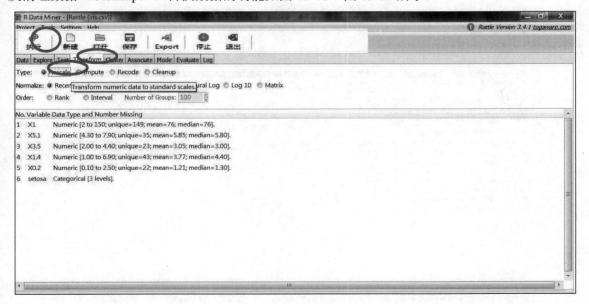

图 13-18 重新缩放

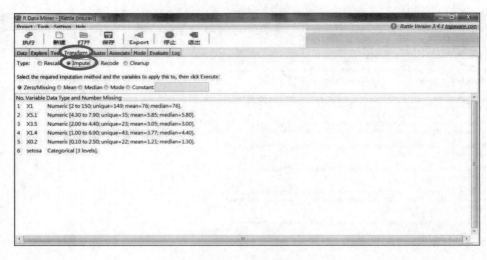

图 13-19 缺失值处理

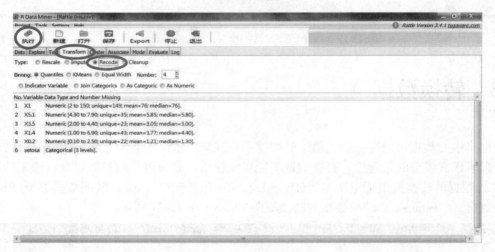

图 13-20 转换数据类型

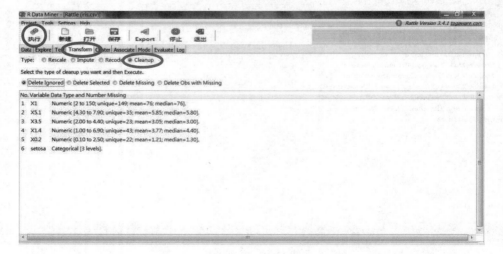

图 13-21 清理数据

13.4 建立、评估及导出模型

经过导入、探索及转换数据（此两者为非必要的步骤）后，我们就可以建立模型了。rattle 提供了聚类（Cluster）、关联性规则（Associate）、分类及回归等算法来建立模型。以 weather.csv 为例，我们可以先导入数据，再建立聚类、关联性法则、决策树及随机森林等模型，其执行界面如图 13-22～图 13-26 所示。

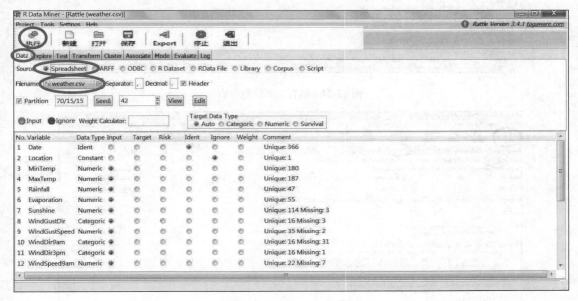

图 13-22 导入 weather.csv

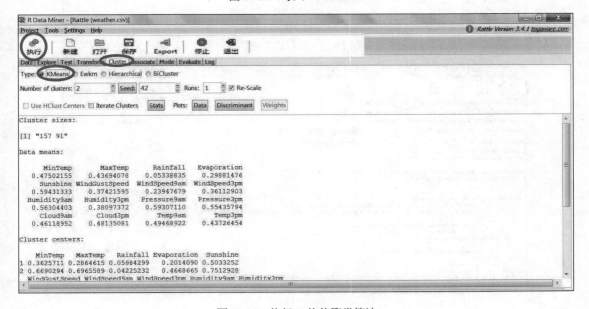

图 13-23 执行 K 均值聚类算法

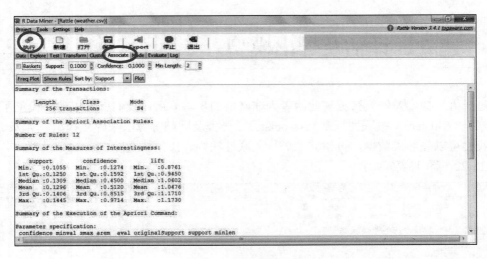

图 13-24　执行关联法则分析

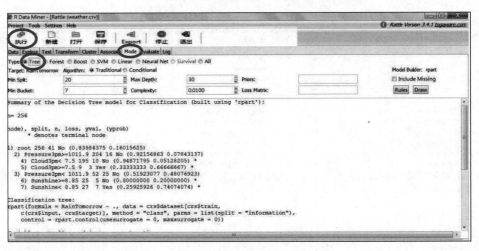

图 13-25　执行决策树算法

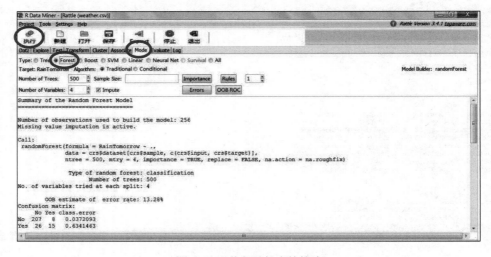

图 13-26　执行随机森林算法

建立模型后，我们就可评估已经建立完成的模型的性能。以 weather.csv 为例，我们可以评估决策树和随机森林的误差矩阵（Error Matrix），其执行界面如图 13-27 所示。我们可以使用 rattle 提供的日志（Log）来查看程序代码，如图 13-28 所示。也可以使用 pmml 包将 R 程序代码导出为 XML 格式的 PMML（Predictive Model Markup Language）。

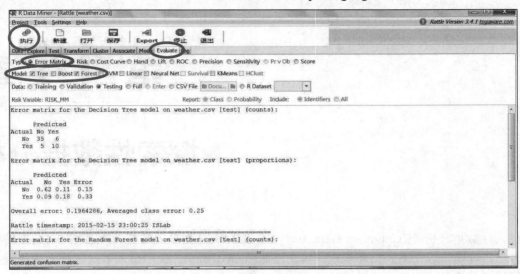

图 13-27　评估决策树和随机森林的误差矩阵

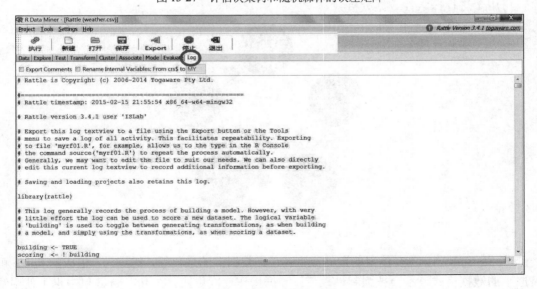

图 13-28　使用日志查看程序代码

13.5　习题

使用 rattle 分析关于银行定期存款的 bank.csv 文档。

第 14 章

探索性数据分析

探索性数据分析（Exploratory Data Analysis，EDA）是指导入数据、数据清理、数据转换、建立算法、可视化等一系列循环流程，其流程如图 14-1 所示。我们必须多次运行上述流程并随时修正信息，才能完成探索性数据分析。R 语言有相当实用的 tidyverse 包，它是多个相关包的集合，可用于进行探索性数据分析。

图 14-1　探索性数据分析循环流程

tibble 是 tidyverse 主要使用的数据类型，它是数据框的进化版，优点为：生成的 tibble 可以保持原来的数据格式，不会被强制性改变；数据操作速度更快。在 tidyverse 包中也可使用管道（pipe）、dplyr 包及 ggplot2 包绘图。

14.1　dplyr 数据处理库（包）

dplyr 处理数据的函数主要有：select()、filter()、arrange()、mutate()、group_by()、n() 及

summarise()。select()函数用于选择特定列（Column）的数据；filter()函数是按照自定义的逻辑条件来筛选出符合条件的数据；arrange()函数用于选择特定列的数据来进行排序；mutate()函数是对现有列的数据进行运算并产生新的列数据；group_by()函数用于选择列数据来进行群组化；n()函数用于统计群组中数据的笔数；summarise()函数用于对数据进行统计运算并返回结果。

[程序范例 14-1]

首先引用 tidyverse 包和 iris 数据集：

```
> library(tidyverse)
> data(iris)
```

调用 select()选择 iris 中的 Petal.Length 和 Petal.Width 的行数据并赋值给 sub_iris 对象：

```
> sub_iris <- select(iris, Petal.Length, Petal.Width)
```

显示 sub_iris 对象前 6 笔数据：

```
> head(sub_iris)
  Petal.Length Petal.Width
1          1.4         0.2
2          1.4         0.2
3          1.3         0.2
4          1.5         0.2
5          1.4         0.2
6          1.7         0.4
```

调用 filter()函数来筛选出符合 Petal.Length <1 或 Petal.Width <1 条件的前 6 笔数据：

```
> head(filter(iris, Petal.Length < 1 | Petal.Width <1))
  Sepal.Length Sepal.Width Petal.Length Petal.Width Species
1          5.1         3.5          1.4         0.2  setosa
2          4.9         3.0          1.4         0.2  setosa
3          4.7         3.2          1.3         0.2  setosa
4          4.6         3.1          1.5         0.2  setosa
5          5.0         3.6          1.4         0.2  setosa
6          5.4         3.9          1.7         0.4  setosa
```

调用 filter()函数来筛选出符合 Petal.Length 为 1.4 且 Petal.Width 为 0.2 条件的前 6 笔数据：

```
> head(filter(iris, Petal.Length == 1.4 & Petal.Width == 0.2))
  Sepal.Length Sepal.Width Petal.Length Petal.Width Species
1          5.1         3.5          1.4         0.2  setosa
2          4.9         3.0          1.4         0.2  setosa
3          5.0         3.6          1.4         0.2  setosa
4          4.4         2.9          1.4         0.2  setosa
5          5.2         3.4          1.4         0.2  setosa
6          5.5         4.2          1.4         0.2  setosa
```

调用 arrange() 函数选择 Petal.Length 进行降序排序并显示前 6 笔数据：

```
> head(arrange(iris, desc(Petal.Length)))
  Sepal.Length Sepal.Width Petal.Length Petal.Width   Species
1          7.7         2.6          6.9         2.3 virginica
2          7.7         3.8          6.7         2.2 virginica
3          7.7         2.8          6.7         2.0 virginica
4          7.6         3.0          6.6         2.1 virginica
5          7.9         3.8          6.4         2.0 virginica
6          7.3         2.9          6.3         1.8 virginica
```

调用 arrange() 函数选择 Petal.Length 进行降序排序、Petal.Width 进行升幂排序并显示前 6 笔数据（以 Petal.Length 先进行降序排序，而后再以 Petal.Width 进行升幂排序）：

```
> head(arrange(iris, desc(Petal.Length), Petal.Width))
  Sepal.Length Sepal.Width Petal.Length Petal.Width   Species
1          7.7         2.6          6.9         2.3 virginica
2          7.7         2.8          6.7         2.0 virginica
3          7.7         3.8          6.7         2.2 virginica
4          7.6         3.0          6.6         2.1 virginica
5          7.9         3.8          6.4         2.0 virginica
6          7.3         2.9          6.3         1.8 virginica
```

调用 arrange() 函数选择 Petal.Width 进行升幂排序、Petal.Length 进行降序排序并显示前 6 笔数据（以 Petal.Width 先进行升幂排序，而后再以 Petal.Length 进行降序排序）：

```
> head(arrange(iris, Petal.Width, desc(Petal.Length)))
  Sepal.Length Sepal.Width Petal.Length Petal.Width Species
1          4.9         3.1          1.5         0.1  setosa
2          5.2         4.1          1.5         0.1  setosa
3          4.8         3.0          1.4         0.1  setosa
4          4.9         3.6          1.4         0.1  setosa
5          4.3         3.0          1.1         0.1  setosa
6          4.8         3.4          1.9         0.2  setosa
```

调用 mutate() 函数产生新 Petal.Length.new 列数据并显示前 6 笔数据：

```
> head(mutate(iris, Petal.Length.new = Petal.Length/ 10))
  Sepal.Length Sepal.Width Petal.Length Petal.Width Species Petal.Length.new
1          5.1         3.5          1.4         0.2  setosa             0.14
2          4.9         3.0          1.4         0.2  setosa             0.14
3          4.7         3.2          1.3         0.2  setosa             0.13
4          4.6         3.1          1.5         0.2  setosa             0.15
5          5.0         3.6          1.4         0.2  setosa             0.14
6          5.4         3.9          1.7         0.4  setosa             0.17
```

调用 group_by() 函数并按照 Petal.Width 列数据来进行群组化并显示前 6 笔数据：

```
> by_Petal.Width <- group_by(iris, Petal.Width)
> head(by_Petal.Width)
Source: local data frame [6 x 5]
Groups: Petal.Width [2]

  Sepal.Length Sepal.Width Petal.Length Petal.Width Species
         <dbl>       <dbl>        <dbl>       <dbl>  <fctr>
1          5.1         3.5          1.4         0.2  setosa
2          4.9         3.0          1.4         0.2  setosa
3          4.7         3.2          1.3         0.2  setosa
4          4.6         3.1          1.5         0.2  setosa
5          5.0         3.6          1.4         0.2  setosa
6          5.4         3.9          1.7         0.4  setosa
```

调用 summarise() 和 n() 函数来计算群组的数量（笔数）及各群组的平均值：

```
> summarise(by_Petal.Width, n=n(), mean(Petal.Width))
# A tibble: 22 × 3
   Petal.Width     n `mean(Petal.Width)`
         <dbl> <int>               <dbl>
1          0.1     5                 0.1
2          0.2    29                 0.2
3          0.3     7                 0.3
4          0.4     7                 0.4
5          0.5     1                 0.5
6          0.6     1                 0.6
7          1.0     7                 1.0
8          1.1     3                 1.1
9          1.2     5                 1.2
10         1.3    13                 1.3
# ... with 12 more rows
```

注意，以上两个函数会将结果转换成 tbl/tibble 的格式，可调用 as.data.frame() 函数把结果转换为数据框。

```
> df_by_Petal.Width <- as.data.frame(by_Petal.Width)
> str(df_by_Petal.Width)
'data.frame':   150 obs. of  5 variables:
 $ Sepal.Length: num  5.1 4.9 4.7 4.6 5 5.4 4.6 5 4.4 4.9 ...
 $ Sepal.Width : num  3.5 3 3.2 3.1 3.6 3.9 3.4 3.4 2.9 3.1 ...
 $ Petal.Length: num  1.4 1.4 1.3 1.5 1.4 1.7 1.4 1.5 1.4 1.5 ...
 $ Petal.Width : num  0.2 0.2 0.2 0.2 0.2 0.4 0.3 0.2 0.2 0.1 ...
 $ Species     : Factor w/ 3 levels "setosa","versicolor",..: 1 1 1 1 1 1 1 1 1 1 ...
```

我们也可以使用 pipe %>%完成上述程序，用法是将左边的 x 作为输入参数赋予到右边的 f 函数再输出结果，如图 14-2 所示。

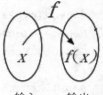

图 14-2　pipe 的作用

```
> sub_iris <- select(iris, Petal.Length, Petal.Width)
> head(sub_iris)
  Petal.Length Petal.Width
1          1.4         0.2
2          1.4         0.2
3          1.3         0.2
4          1.5         0.2
5          1.4         0.2
6          1.7         0.4
> iris %>%
+   select(Petal.Length, Petal.Width) %>%
+   head()
  Petal.Length Petal.Width
1          1.4         0.2
2          1.4         0.2
3          1.3         0.2
4          1.5         0.2
5          1.4         0.2
6          1.7         0.4
```

ggplot()函数是以一种图层式的概念在绘图，每一个图层上的数据可以有不同的来源和映射（Mapping）。映射是指数据中的各个变量如何与图形上的各种属性相对应。ggplot()函数的基本要素包含数据（Data）和映射（Mapping）、几何对象（Geometric）、标尺（Scale）、统计（Statistics）、坐标系统（Coordinate）、图层（Layer）、分面（Facet）和主题（Theme）。另外，qplot()函数只允许单个数据来源以及一组映射。

[程序范例 14-2]

可调用 qplot()函数来显示图形：

```
> library(tidyverse)
```

使用 iris 数据集，以 Sepal.Length 为 X 轴、Sepal.Width 为 Y 轴并以颜色显示 Species：

```
# x轴= Sepal.Length, y轴=Sepal.Widt
> qplot(data=iris, Sepal.Length, Sepal.Width, colour=Species)
```

结果如图 14-3 所示。

以 Sepal.Length 为 X 轴、Sepal.Width 为 Y 轴并以形状显示 species：

```
> qplot(data=iris, Sepal.Length, Sepal.Width, shape=Species)
```

结果如图 14-4 所示。

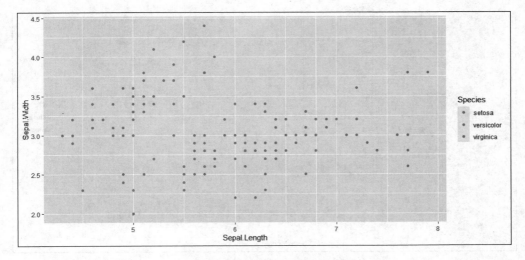

图 14-3　调用 qplot()H 函数以不同颜色显示 Species

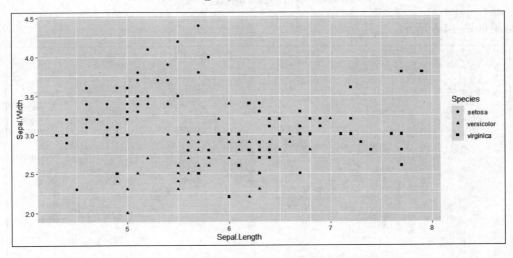

图 14-4　调用 qplot()函数以不同颜色显示 Species

调用 ggplot()函数时，aes 是指映射，可以调用 geom_point()绘制散点图。

```
> iris1 <- iris %>%
+   filter(Sepal.Length > 5)
> ggplot(iris1, aes(x = Sepal.Length,
+            y = Sepal.Width,
+            color=Species)) + geom_point()
```

结果如图 14-5 所示。

可以在 aes()映射中增加有关尺寸的参数：

```
> ggplot(iris1, aes(x = Sepal.Length,
+            y = Sepal.Width,
+            color=Species,
+            size=Petal.Length)) + geom_point()
```

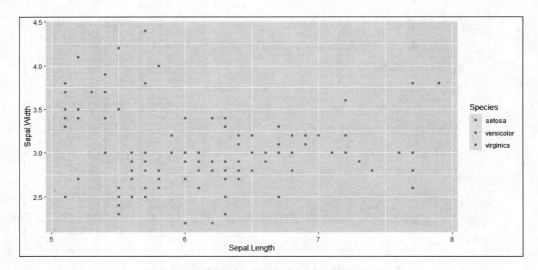

图 14-5　调用 ggplot()函数绘制散点图

结果如图 14-6 所示。

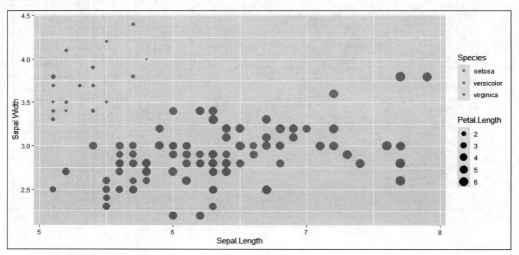

图 14-6　调用 ggplot()函数绘制散点图并增加映射参数

分面可以按照特定的条件对数据进行分组，然后分别绘图：

```
> ggplot(iris1, aes(x = Sepal.Length,
+                   y = Sepal.Width,
+                   size=Petal.Length)) +
+     geom_point() +
+     facet_wrap(~Species)
```

结果如图 14-7 所示。

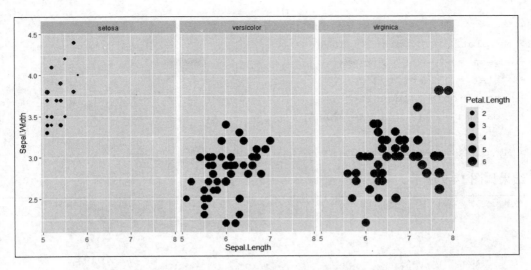

图 14-7　调用 ggplot()函数绘制散点图并按 Species 分成 3 组

绘制线可调用 geom_line()函数，绘制条形图可调用 geom_col()函数，绘制直方图可调用 geom_histogram()函数，绘制盒形图可调用 geom_boxplot()函数，诸如此类。

```
> ggplot(iris1, aes(x = Sepal.Length,
+                   y = Sepal.Width)) + geom_line()
```

结果如图 14-8 所示。

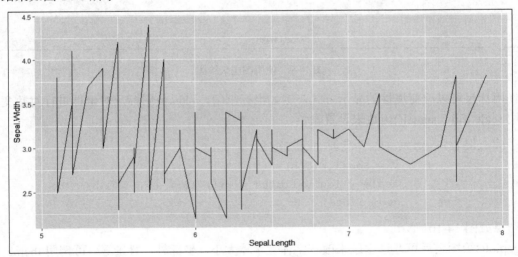

图 14-8　调用 ggplot()函数绘制线

我们也可以使用图层来绘图，其中 my.plot1 尚未包含任何图层，所以还无法显示出任何图形。My.plot2 在 My.plot1 创建的绘图对象上再加入一个 geom 指定为 point 的图层，即可产生一张散点图，其中 stat 与 position 参数分别用来指定统计方法与调整几何图形的位置，若不需要特别地指定与调整，则指定为 identity。params 则是用来指定 geom 与 stat 所需要的参数。

```
> My.plot1=ggplot(iris1, aes(x = Sepal.Length,
+                            y = Sepal.Width,
```

```
+                         color=Species))
> My.plot2 <- My.plot1 + layer (
+   geom =  "point" ,
+   stat =  "identity" ,
+   position =  "identity" ,
+   params =  list(na.rm= FALSE)
+ )
> My.plot2
```

结果如图 14-9 所示。

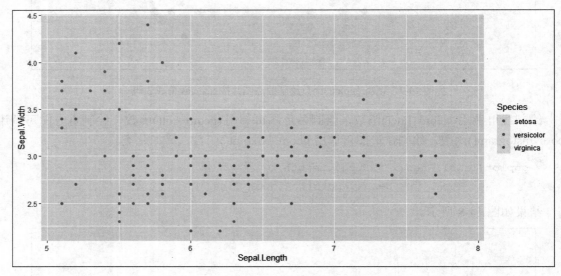

图 14-9　使用图层来绘图

使用标尺可以改变属性的显示方式，例如坐标刻度，可以使用标尺对坐标进行对数变换，或调用 scale_fill_manual()函数来设置颜色。

```
> ggplot(iris1, aes(x = Sepal.Length,
+                   y = Sepal.Width)) +
+         scale_y_log10(limits = c(1, 10)) +
+         geom_line()
```

结果如图 14-10 所示。

ggplot()绘图之后可通过主题来改变字体、字体大小、坐标轴、背景等。可调用 theme_bw()函数改为白色背景的主题或调用 theme_classic()函数改为传统的主题。

```
> ggplot(iris1, aes(x = Sepal.Length,
+                   y = Sepal.Width)) +
+         scale_y_log10(limits = c(1, 10)) +
+         geom_line()+
+         theme_bw()
```

结果如图 14-11 所示。

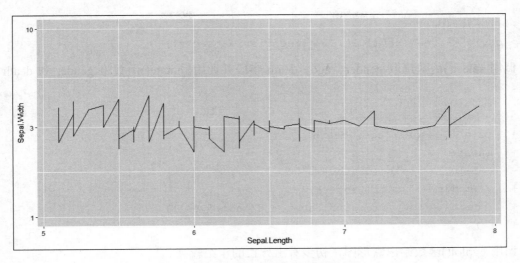

图 14-10　使用标尺对坐标进行对数变换

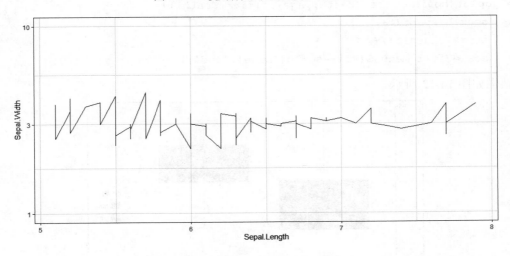

图 14-11　调用 theme_bw()函数改为白色背景的主题

14.2　案例分析

本节以对实际案例的分析来综合说明探索性数据分析。

[程序范例 14-3]　新冠肺炎分析案例

可以到网址 https://www.kaggle.com/sudalairajkumar/novel-corona-virus-2019-dataset 下载 COVID19_line_list_data.csv。

```
> library(tidyverse)
> library(rpart)
> library("e1071")
```

```
> library(party)
> covid <- read.csv('COVID19_line_list_data.csv')
```

调用 select()函数选取 gender、age、death 字段并调用 mutate()函数将 gender 和 death 转成因子：

```
> covid_1 <- covid %>%
+   select(gender, age, death) %>%
+   mutate(
+     death = ifelse(death == '0', 0, 1),
+     gender = factor(gender),
+     death = factor(death, label = c('no','yes'))
+   )
```

调用 **ggplot**()函数绘出盒形图，初步看出死亡的年纪较高。

```
> ggplot(covid_1, aes(death, age, fill = death))+
+   geom_boxplot()+
+   theme_classic()+
+   scale_fill_manual(values = c('blue','red'))
```

结果如图 14-12 所示。

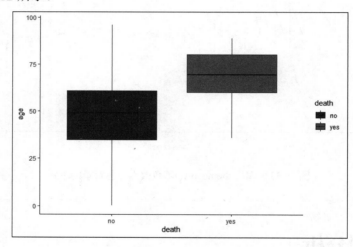

图 14-12　绘出盒形图

调用 select()函数选取 gender、age、death 字段并调用 mutate()函数将 gender 和 age 转成数值、death 转成因子。

```
> covid_2 <- covid %>%
+   select(gender, age, death) %>%
+   mutate(
+     gender = ifelse(gender== 'male',0,1),
+     gender = as.numeric(gender),
+     age = as.numeric(age),
+     death = ifelse(death == '0', 0, 1),
```

```
+     death = factor(death, label = c('no','yes'))
+   )
```

将数据集分为训练和测试数据（300 笔）。

```
> covid_2= na.omit(covid_2)
> covid_3=covid_2
> test.index = sample(1:nrow(covid_3),300)
> covid_new_test = covid_3[test.index,]         # Testing
> covid_new_train = covid_3[-test.index,]       # Training
```

调用 rpart()函数来测试 300 笔测试集。

```
> seed(40)
> model_tree = rpart(death ~ age + gender, method="class",
+ data= covid_new_train, control=rpart.control(minsplit=5))
> rpart.plot(model_tree, type = 5)
```

结果如图 14-13 所示。

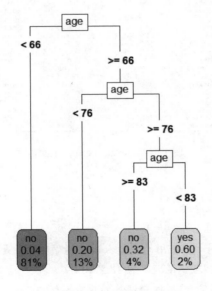

图 14-13　决策树规则

可得到 4 条决策树规则（我们可重新设置 seed 和 minsplit 参数，产生不同规则）：

（1）age< 65.5 427 17 no *
（2）age>=65.5 and age< 75.5 no   *
（3）age>=75.5 and age>=82.5   no   *
（4）age>=75.5 and age< 82.5   yes   *

```
> test.predict=factor(predict(model_tree,covid_new_test,
+ type='class'), levels=levels(covid_new_test$death))
> table.testdata =table(covid_new_test$death,test.predict)
> table.testdata
```

```
         test.predict
          no  yes
     no  279    6
     yes  11    4
```

准确率=（279+4）/（279+4+11+6）=0.943。

[程序范例 14-4] 美国医疗保险费用案例

分析美国医疗保险费用案例，使用 insurance.csv 数据集，读者可使用本书提供的 insurance.csv，数据集有 6 个变量，如图 14-14 所示。

	A	B	C	D	E	F	G
1	age	sex	bmi	children	smoker	region	charges
2	19	female	27.9	0	yes	southwest	16884.924
3	18	male	33.77	1	no	southeast	1725.5523
4	28	male	33	3	no	southeast	4449.462
5	33	male	22.705	0	no	northwest	21984.471
6	32	male	28.88	0	no	northwest	3866.8552
7	31	female	25.74	0	no	southeast	3756.6216
8	46	female	33.44	1	no	southeast	8240.5896
9	37	female	27.74	3	no	northwest	7281.5056
10	37	male	29.83	2	no	northeast	6406.4107
11	60	female	25.84	0	no	northwest	28923.137
12	25	male	26.22	0	no	northeast	2721.3208
13	62	female	26.29	0	yes	southeast	27808.725
14	23	male	34.4	0	no	southwest	1826.843
15	56	female	39.82	0	no	southeast	11090.718
16	27	male	42.13	0	yes	southeast	39611.758
17	19	male	24.6	1	no	southwest	1837.237
18	52	female	30.78	1	no	northeast	10797.336
19	23	male	23.845	0	no	northeast	2395.1716
20	56	male	40.3	0	no	southwest	10602.385
21	30	male	35.3	0	yes	southwest	36837.467
22	60	female	36.005	0	no	northeast	13228.847
23	30	female	32.4	1	no	southwest	4149.736

图 14-14 insurance.csv 数据集

其中：

- age（年龄）：主要受益人年龄。
- sex（性别）：受益人性别（男、女）。
- bmi（身体质量指数）：国际上常用的衡量人体胖瘦程度以及是否健康的一个标准，计算公式为体重与身高的平方之比，理想情况下为 18.5～24.9。
- children（儿童）：抚养人数/健康保险涵盖的儿童人数。
- smoker（吸烟者）：表示受益人是否吸烟。
- region（地区）：表示受益人居住在美国的地区。
- charges（费用）：健康保险支付的个人医疗费用。

```
> library(rattle)         # Access the weather dataset and utilities.
> library(magrittr)       # Utilise %>% and %<>% pipeline operators.
> library(e1071)
> library(rpart)
> library(rpart.plot)
> library(ggplot2)
> library(randomForest)
```

```
> library(Hmisc, quietly=TRUE)

> building <- TRUE
> scoring  <- ! building

# 设置随机数的种子数，结果可重复操作
> crv$seed <- 42
> set.seed(crv$seed)

# 读取数据集
> fname        <- "file:///C:/Temp/insurance.csv"
> crs$dataset <- read.csv(fname, encoding="UTF-8")

# 设置70%为训练集、30%为测试集
# nobs=1338 train=936 test=402

> crs$nobs <- nrow(crs$dataset)
> crs$train <- sample(crs$nobs, 0.7*crs$nobs)
> crs$validate <- NULL

> crs$nobs %>%
>  seq_len() %>%
>  setdiff(crs$train) %>%
>  setdiff(crs$validate) ->
>  crs$test

> length(crs$train)
> length(crs$test)

# 设置输入、输出变量和类型
> crs$input     <- c("age", "sex", "bmi", "children", "smoker","region")
> crs$numeric   <- c("age", "bmi", "children")
> crs$categoric <- c("sex", "smoker", "region")
> crs$target    <- "charges"
> crs$risk      <- NULL
> crs$ident     <- NULL
> crs$ignore    <- NULL
> crs$weights   <- NULL

# 调用Hmisc库的函数显示数据基本统计信息
> contents(crs$dataset[crs$train, c(crs$input, crs$risk, crs$target)])
> summary(crs$dataset[crs$train, c(crs$input, crs$risk, crs$target)])

       age            sex              bmi          children        smoker
 Min.   :18.00   Length:936       Min.   :17.29   Min.   :0.000   Length:936
```

```
         1st Qu.:27.00   Class :character   1st Qu.:26.41   1st Qu.:0.000
 Class :character
         Median :40.00   Mode  :character   Median :30.69   Median :1.000
 Mode  :character
         Mean   :39.61                      Mean   :30.87   Mean   :1.107
         3rd Qu.:52.00                      3rd Qu.:34.87   3rd Qu.:2.000
         Max.   :64.00                      Max.   :53.13   Max.   :5.000

     region              charges
 Length:936          Min.   : 1136
 Class :character    1st Qu.: 4845
 Mode  :character    Median : 9675
                     Mean   :13365
                     3rd Qu.:16580
                     Max.   :63770
```

我们可以使用盒形图判断地区、吸烟、性别、儿童和身体质量指数对医疗保险费用的关系。地区与医疗保险费用的关系：

```
# Charges vs region, visualization
> ggplot(data = crs$dataset, aes(x=region, y=charges)) +
>   geom_boxplot(fill = c(6:9))+
>   ggtitle("Medical charges per region")
```

从图 14-15 可知，美国所有地区的医疗保险费用几乎是一样的。

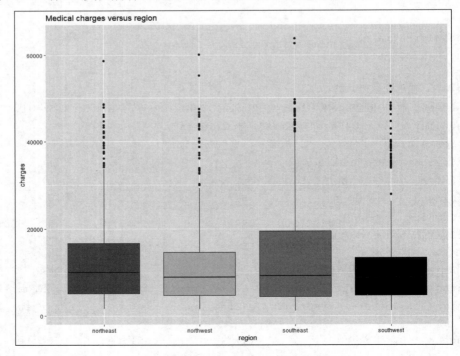

图 14-15　地区对医疗保险费用的盒形图

吸烟与医疗保险费用的关系：

```
# Charges vs smoking
charge_mean <- function(x){
  return (round((data.frame(y=mean(x),label=mean(x,na.rm=T))),2))}

ggplot(data = crs$dataset, aes(x=smoker, y=charges)) +
  geom_boxplot(fill = c(6:7))+
  stat_summary(fun.y = mean,geom="point",colour="darkred",size=3) +
  stat_summary(fun.data = charge_mean,geom="text",vjust=-0.7) +
  ggtitle("Medical charges versus smoking")
```

从图 14-16 所示的吸烟与医疗保险费用关系的盒形图可知，与非吸烟者相比，吸烟者的医疗保险费用（32050.23）平均值很高，几乎是非吸烟者费用的 4 倍。盒形图上的圆点代表医疗保险费用的平均值。

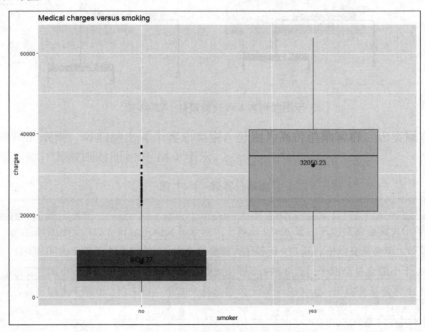

图 14-16　吸烟与医疗保险费用关系的盒形图

性别与医疗保险费用的关系：

```
# Medical Charges based on gender
> ggplot(data = crs$dataset, aes(x=sex, y=charges)) +
>   geom_boxplot(fill = c(6:7))+
>   stat_summary(fun.y=mean, geom="point",colour="darkred", size=3) +
>   stat_summary(fun.data = charge_mean, geom="text", vjust=-0.8) +
>   ggtitle("Medical charges versus gender")
```

从图 14-17 所示的性别与医疗保险费用关系的盒形图可知，男性和女性的医疗保险费用几乎是相同的，与性别特征无关。

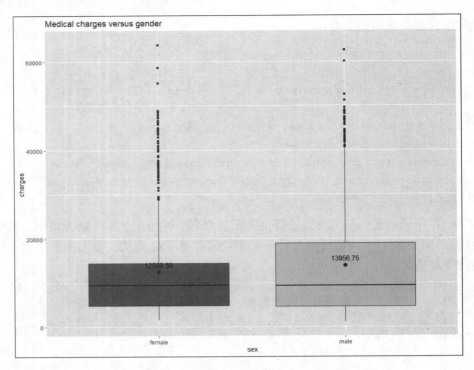

图 14-17　性别与医疗保险费用关系的盒形图

孩子的数量与医疗保险费用的关系：

```
# Medical Charges based on children
ggplot(data = crs$dataset, aes(x=factor(children), y=charges)) +
  geom_boxplot(aes(fill = children))+
  stat_summary(fun = mean, geom="point",colour="darkred", size=3) +
  stat_summary(fun.data = charge_mean, geom="text", vjust=-0.8) +
  ggtitle("Medical charges versus children")
```

从图 14-18 所示孩子的数量与医疗保险费用关系的盒形图可知，有两个孩子的医疗保险费用较高，5 个孩子的医疗保险费用较低。

肥胖与医疗保险费用的关系，肥胖者是以 BMI > 30 来认定的：

```
#Medical cost by BMI
crs$dataset$obesity <- ifelse(crs$dataset$bmi > 30, "yes", "no")
head(crs$dataset$obesity, n=2)

ggplot(data = crs$dataset, aes(x=factor(obesity), y=charges)) +
  geom_boxplot(aes(fill = obesity))+
  stat_summary(fun = mean, geom="point",colour="darkred", size=3) +
  stat_summary(fun.data = charge_mean, geom="text", vjust=-0.8) +
  ggtitle("Medical charges versus obesity")
```

从图 14-19 所示肥胖医疗保险费用关系的盒形图可知，对于肥胖者来说，医疗保险费用比不是肥胖者要高出近 50%。

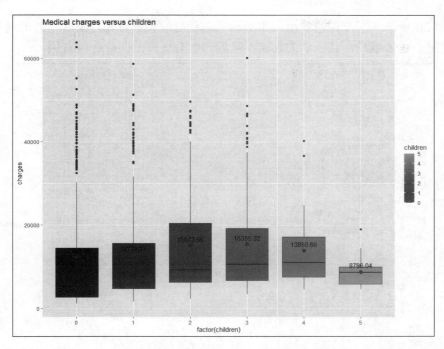

图 14-18　孩子的数量与医疗保险费用关系的盒形图

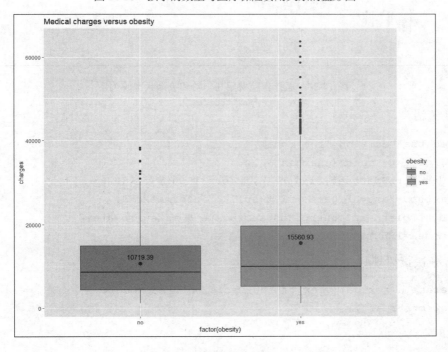

图 14-19　肥胖与医疗保险费用关系的盒形图

医疗保险费用与年龄和吸烟状况的关系：

```
#plot(age,charges,col=smoker)
> ggplot(data=crs$dataset, aes(x=age, y=charges)) +
> geom_point(aes(colour = factor(smoker)))
```

从图 14-20 所示的医疗保险费用与年龄和吸烟状况的关系可知，随着年龄的增长，人们的医疗保险费用会越来越高，且吸烟的人比不吸烟的人有更高的医疗保险费用。

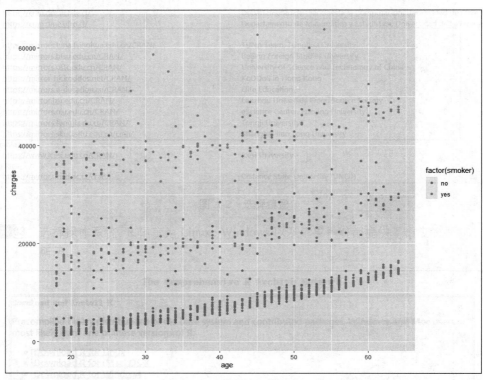

图 14-20　医疗保险费用与年龄和吸烟状况的关系

使用决策树：

```
# Build the Decision Tree model.
> crs$rpart <- rpart(charges ~ .,
+     data=crs$dataset[crs$train, c(crs$input, crs$target)],
+     method="anova",parms=list(split="information"),
+     control=rpart.control(usesurrogate=0,maxsurrogate=0),
+     model=TRUE)
```

产生 5 条决策树规则：

```
# Generate a textual view of the Decision Tree model.
> print(crs$rpart)

n= 936
node), split, n, deviance, yval
      * denotes terminal node

 1) root 936 137095400000 13365.130
   2) smoker=no 747  27880480000  8630.720
     4) age< 42.5 400   8342459000  5333.457 *
```

```
  5) age>=42.5 347  10176250000 12431.600 *
  3) smoker=yes 189 26293480000 32077.340
    6) bmi< 30.1 89   2047455000 21078.980 *
    7) bmi>=30.1 100  3898677000 41865.880
     14) age< 41.5 56  1272613000 38534.450 *
     15) age>=41.5 44  1213537000 46105.880 *
```

决策树图：

```
> rpart.plot(crs$rpart,type=0)
```

结果如图 14-21 所示。

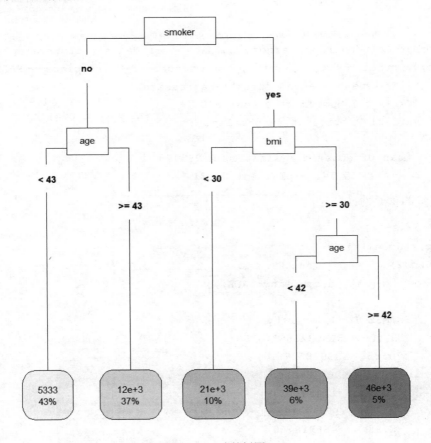

图 14-21　决策树图

预测测试数据集和计算 RMSE：

```
# Predict value
> crs$pr <- predict(crs$rpart, newdata=crs$dataset[crs$test,
+ c(crs$input)])
> rmse=sqrt(sum((crs$pr - crs$test)^2)/length(crs$test))
> print(paste0("Decision tree RMSE: ", round(rmse,2)))
[1] "Decision tree RMSE: 16855.77"
```

使用随机森林分析:

```
#create random forest model
> crs$rf <- randomForest(charges ~ .,
+                        data=crs$dataset[crs$train, c(crs$input,
+                        crs$target)],
+                        ntree=500,
+                        mtry=2,
+                        importance=TRUE,
+                        replace=FALSE)
> crs$rf

Call:
 randomForest(formula = charges ~ ., data = crs$dataset[crs$train,c(crs$input,crs$target)], ntree = 500, mtry = 2, importance = TRUE, replace = FALSE)
               Type of random forest: regression
                     Number of trees: 500
No. of variables tried at each split: 2

          Mean of squared residuals: 22985473
                    % Var explained: 84.31
```

计算变量的重要性:

```
> n <- crs$rf %>%
+   importance() %>%
+   round(2)
> rn[order(rn[,1], decreasing=TRUE),]

         %IncMSE IncNodePurity
smoker    151.05   51353266834
age        70.85   10987855761
bmi        62.48   11671008665
children   13.83    1745717172
region      5.77    1194991426
sex        -2.98     486918760
```

由以上可知,smoker、age 和 bmi 是排名前三的重要变量。

绘出随机森林误差图:

```
> plot(crs$rf, main="")
> legend("topright", c(""), text.col=1:6, lty=1:3, col=1:3)
> title(main="Random Forest")
```

结果如图 14-22 所示。

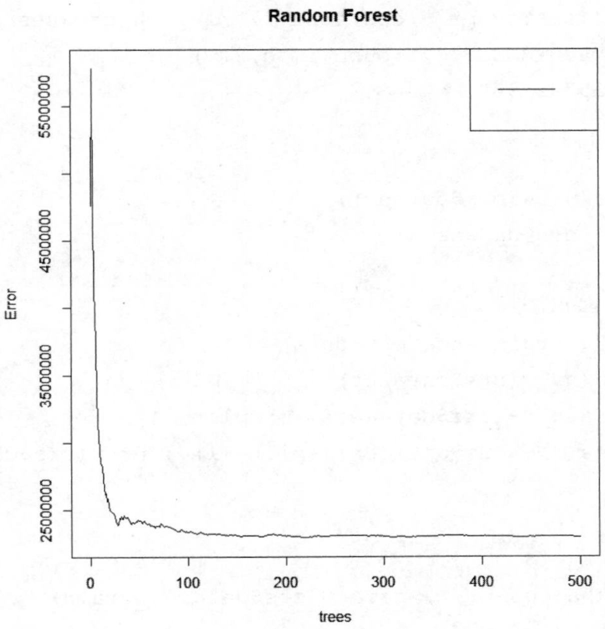

Random Forest

图 14-22 随机森林误差图

使用 SVM 分析：

```
> crs$svm <- svm(charges ~ .,
+         data=crs$dataset[crs$train, c(crs$input, crs$target)],
+         type="nu-regression",
+         decision.values = TRUE
+ )
>
> print(crs$svm)


Call:
svm(formula = charges ~ ., data = crs$dataset[crs$train, c(crs$input,
crs$target)],
    type = "nu-regression", decision.values = TRUE)



Parameters:
   SVM-Type:  nu-regression
   SVM-Kernel:  radial
       cost:  1
         nu:  0.5
```

预测测试数据集和计算 RMSE：

```
> crs$svmpr <- predict(crs$svm, newdata=crs$dataset[crs$test,
+           c(crs$input)])
```

```
> rmse=sqrt(sum((crs$svmpr - crs$test)^2)/length(crs$test))
> print(paste0("SVM RMSE: ", round(rmse,2)))
[1] "SVM RMSE: 15702.96"
```

重新抽样：

```
> rows <- sample(nrow(crs$dataset))
> crs$dataset <- crs$dataset[rows, ]

# 70%为训练集，30%为测试集
# Split the 70/30 train and test data
> split <- round(nrow(crs$dataset) * 0.7 )
> crs$dataset_train <- crs$dataset[1 : split, ]
> crs$dataset_test <- crs$dataset[(split +1) : nrow(crs$dataset),]
```

建立线性回归模型：

```
#Create Model
> lm.fit <- lm(charges ~ . , data = crs$dataset_train)
> summary(lm.fit)

Call:
lm(formula = charges ~ ., data = insurance_train)

Residuals:
   Min     1Q Median     3Q    Max
-11915  -2861   -936   1366  25195

Coefficients:
                  Estimate Std. Error t value Pr(>|t|)
(Intercept)      -11825.99    1189.73  -9.940   <2e-16 ***
age                 253.74      14.17  17.911   <2e-16 ***
sexmale             -17.72     399.19  -0.044   0.9646
bmi                 337.02      34.22   9.850   <2e-16 ***
children            375.10     163.09   2.300   0.0217 *
smokeryes         24122.00     496.06  48.627   <2e-16 ***
regionnorthwest    -180.28     571.95  -0.315   0.7527
regionsoutheast    -797.99     574.07  -1.390   0.1648
regionsouthwest    -688.89     573.75  -1.201   0.2302
---
Signif. codes:  0 '***' 0.001 '**' 0.01 '*' 0.05 '.' 0.1 ' ' 1

Residual standard error: 6059 on 928 degrees of freedom
Multiple R-squared:  0.753, Adjusted R-squared:  0.7509
F-statistic: 353.7 on 8 and 928 DF,  p-value: < 2.2e-16
```

训练集 R-squared 值为 0.753，由于 R-squared 值越接近 1，模型拟合程度越好，我们可以使用合成变量（bmi*smoker*age）再优化。

```
> lm.fit2 <- lm(charges ~ age +sex + bmi + children + smoker + region + bmi * smoker *age, data=insurance_train)
> summary(lm.fit2)

Call:
lm(formula = charges ~ age + sex + bmi + children + smoker +
    region + bmi * smoker * age, data = insurance_train)

Residuals:
     Min      1Q  Median      3Q     Max
-10643.7 -1771.0 -1258.7  -466.8 30842.2

Coefficients:
                   Estimate Std. Error t value Pr(>|t|)
(Intercept)       -1552.436   2718.763  -0.571 0.568133
age                 229.324     67.234   3.411 0.000676 ***
sexmale            -560.977    320.192  -1.752 0.080105 .
bmi                 -22.990     88.130  -0.261 0.794252
children            432.609    132.612   3.262 0.001146 **
smokeryes        -14057.443   6110.710  -2.300 0.021644 *
regionnorthwest    -126.384    451.886  -0.280 0.779785
regionsoutheast   -1101.930    458.251  -2.405 0.016384 *
regionsouthwest    -648.878    454.593  -1.427 0.153807
bmi:smokeryes      1268.942    196.684   6.452 1.78e-10 ***
age:bmi               1.503      2.152   0.698 0.485121
age:smokeryes      -153.643    153.923  -0.998 0.318454
age:bmi:smokeryes     4.200      4.920   0.854 0.393578
---
Signif. codes:  0 '***' 0.001 '**' 0.01 '*' 0.05 '.' 0.1 ' ' 1

Residual standard error: 4840 on 924 degrees of freedom
Multiple R-squared:  0.8415, Adjusted R-squared:  0.8395
F-statistic: 408.9 on 12 and 924 DF,  p-value: < 2.2e-16
```

训练集 R-squared 的值变为 0.8415。
预估测试集的医疗保险费用并计算 R-squared 值：

```
> insurance_test$predict <- predict(lm.fit2, newdata=insurance_test)
> rss <- sum((insurance_test$predict - insurance_test$charges)^2)
> tss <- sum((insurance_test$charges - mean(insurance_test$charges))^2)
> (rsq = 1- (rss/tss))
[1] 0.8384732
```

由以上可知，测试集 R-squared 的值为 0.8384732。

计算测试集医疗保险费用 RMSE：

```
> rmse=sqrt(sum((insurance_test$predict -
+ insurance_test$charges)^2)/nrow(insurance_test))
> rmse
[1] 4894.962
> print(paste0("Linear model RMSE: ", round(rmse,2)))
[1] "Linear model RMSE: 4894.96"
```

绘出测试集的预估医疗保险费用，如图 14-23 所示。

```
> #Visualization of prediction
> ggplot(data=insurance_test, aes(x=predict, y=charges))+
+   geom_point()+
+   geom_abline()
```

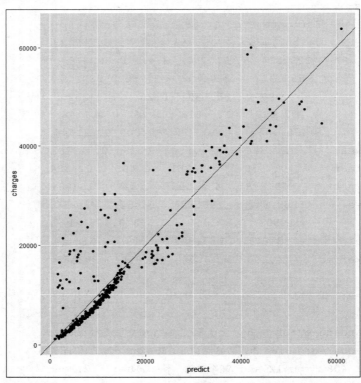

图 14-23　预估测试集的医疗保险费用图

预估年轻男性的医疗保险费用：

```
> guy <- data.frame(age = 15,
+           sex="male",
+           bmi = 27.9,
+           children = 0,
```

```
+               smoker = "no",
+               region = "northwest")
> print(paste0("Medical charges for a teen ager: ",
+ round(predict(lm.fit2, guy), 2)))
[1] "Medical charges for a teen ager: 1187.53"
```

预估年长男性的医疗保险费用：

```
> man <- data.frame(age = 70,
+                   sex="male",
+                   bmi = 35,
+                   children = 2,
+                   smoker = "yes",
+                   region = "northwest")
> print(paste0("Medical charges for a man: ", round(predict(lm.fit2, man), 2)))
[1] "Medical charges for a man: 47445.09"
```

[程序范例 14-5] 超市顾客客群分析案例

本案例研究超市营销分析，以案例中提供的公司会员基本数据、交易数据等信息作为分析数据来源，并结合数据挖掘技术的聚类分析深入了解案例中公司的交易类型，进而找出有助于案例中公司的决策信息。掌握不同的 4 项客群，提供各自专门的营销策略，通过针对性营销锁定不同顾客喜爱的商品，提高消费者的忠诚度，以达到最有效的营销模式。本案例采用 RFM 模型，说明如下：

- 最近购买日（Recency）：顾客最近一次购买到分析时的天数，计算数值越小，表示顾客近期有到超级市场购买商品，反之，数值越大，表示顾客越长时间没有到超市购买商品。
- 购买频率（Frequency）：顾客在 2009 年 1 月 1 日至 2010 年 12 月 31 日期间购买超市商品的次数，计算数值越大，表示顾客越常购买，反之，数值越小，表示顾客越不常购买。
- 购买金额（Monetary）：顾客在 2017 年 1 月 1 日至 2018 年 12 月 31 日期间购买超市商品的平均总金额，计算数值越大，表示顾客消费能力越高，反之，计算数值越小，表示顾客消费能力越低。

首先引用相关包（或库）：

```
> library(magrittr) # Utilise %>% and %<>% pipeline operators.
> library(e1071)
> library(rpart)
> library(rpart.plot)
> library(ggplot2)
> library(randomForest)
> library(NbClust)
```

由于本案例采用 RFM 模型，先调用 getRFM 函数计算 RFM 值。

```
getRFM <- function(df,startDate,endDate,,tIDColName="ID",tDateColName="Date",
tAmountColName="Amount")
```

其中：

- df：包含客户 ID、日期和交易金额的数据框。
- startDate：交易的开始日期，发生的交易在开始日期之后将保留。
- endDate：交易的结束日期，与开始日期一起使用来设置时间范围。
- tIDCOLName：在输入数据框中包含客户 ID 的列名称。
- tDateColName：在输入数据框中包含日期的列名称。
- tamountcolname：在输入数据框中包含交易金额的列名称。

读取 super.csv 并统计基本数据：

```
> aaa=read.csv("c:/Temp/super.csv", header=T, sep="," ,
            stringsAsFactors=F)
> name=c("shop","ID","Date","Amount")
> names(aaa) <- name
> str(aaa)
> summary(aaa)
      shop           ID            Date              Amount
 Min.   :1023   Min.   :   1   Length:12969       Min.   : -149.0
 1st Qu.:1023   1st Qu.: 856   Class :character   1st Qu.:  349.0
 Median :1051   Median :1814   Mode  :character   Median :  571.0
 Mean   :1206   Mean   :1907                      Mean   :  763.6
 3rd Qu.:1069   3rd Qu.:2824                      3rd Qu.:  936.0
 Max.   :4002   Max.   :4377                      Max.   :12631.0
```

调用 ggplot()函数显示超市编号、客户 ID 和交易费用的数据：

```
> ggplot(data=aaa, aes(x=shop, y=sum(Amount), color=ID))+
+   geom_count()
```

结果如图 14-24 所示。

删除不用的字段：

```
> a=aaa[,-1] #Remove the shop field
> str(a)
'data.frame':   12969 obs. of  3 variables:
 $ ID    : int  1 1 2 2 3 3 4 5 5 5 ...
 $ Date  : chr  "2009/5/29" "2009/10/25" "2009/9/11" "2009/9/16" ...
 $ Amount: int  199 1488 389 346 495 296 370 424 351 368 ...
```

转换日期格式：

```
> a$Date = as.Date(a$Date,"%Y/%m/%d")
> head(a)
  ID       Date Amount
1  1 2009-05-29    199
2  1 2009-10-25   1488
3  2 2009-09-11    389
```

```
4  2 2009-09-16     346
5  3 2009-04-11     495
6  3 2009-04-10     296
```

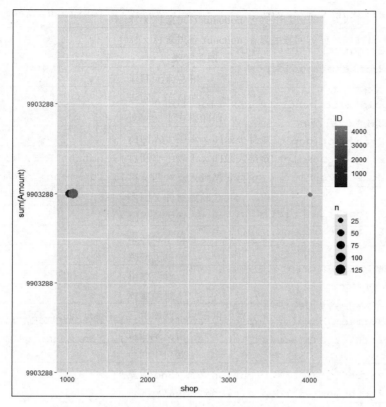

图 14-24　超市编号、客户 ID 和交易费用图

设置交易的开始日期和结束日期：

```
> startDate <- as.Date("20090101","%Y%m%d")
> endDate <- as.Date("20101231","%Y%m%d")
```

调用 getRFM() 函数产生 Recency、Frequency、Monetary 数据：

```
> RFM <- getRFM(a,startDate,endDate)
> RFM_Model=RFM[, 4:6]
> summary(RFM_Model)
    Recency         Frequency         Monetary
 Min.   :  0.0    Min.   :  1.000   Min.   :   0.0
 1st Qu.: 77.0    1st Qu.:  1.000   1st Qu.: 369.3
 Median :236.0    Median :  1.000   Median : 587.0
 Mean   :280.1    Mean   :  2.963   Mean   : 744.8
 3rd Qu.:474.0    3rd Qu.:  3.000   3rd Qu.: 919.0
 Max.   :729.0    Max.   :125.000   Max.   :9521.5
```

调用 ggplot() 函数显示 Recency、Frequency、Monetary 数据：

```
> p01 <- RFM_Model %>%
+   ggplot2::ggplot(ggplot2::aes(y=Recency)) +
+   ggplot2::geom_boxplot(ggplot2::aes(x="All"), notch=TRUE, fill="grey") +
+   ggplot2::stat_summary(ggplot2::aes(x="All"), fun=mean, geom="point", shape=8) +
+   ggplot2::ggtitle("Distribution of Recency ") +
+   ggplot2::theme(legend.position="none")
>
> # Use ggplot2 to generate box plot for F
> p02 <- RFM_Model %>%
+   ggplot2::ggplot(ggplot2::aes(y=Frequency)) +
+   ggplot2::geom_boxplot(ggplot2::aes(x="All"), notch=TRUE, fill="grey") +
+   ggplot2::stat_summary(ggplot2::aes(x="All"), fun=mean, geom="point", shape=8) +
+   ggplot2::ggtitle("Distribution of Frequency") +
+   ggplot2::theme(legend.position="none")
>
> # Use ggplot2 to generate box plot for M
> p03 <- RFM_Model %>%
+   ggplot2::ggplot(ggplot2::aes(y=Monetary)) +
+   ggplot2::geom_boxplot(ggplot2::aes(x="All"), notch=TRUE, fill="grey") +
+   ggplot2::stat_summary(ggplot2::aes(x="All"), fun=mean, geom="point", shape=8) +
+   ggplot2::ggtitle("Distribution of Monetary") +
+   ggplot2::theme(legend.position="none")
> gridExtra::grid.arrange(p01, p02, p03)
```

结果如图14-25所示。

先评估调用kmeans()函数（k均值聚类算法）分客群为k=1:6群时的结果。从图14-25中可以明显看出，k在4之后减小的幅度变缓，这说明在k=4之后，如果增加聚类的类别，效果提高不是特别明显，由此可以看出可聚为4类。

```
> wss<- 0
> for(i in 1:6) {
+   model_km= kmeans(RFM_Model, i, nstart=20)
+   wss[i]=model_km$tot.withinss
+ }
> plot(RFM_Model, col = model_km$cluster,
+      main = "Model Sum of Squares")
>
> plot(1:6, wss, type = "b",
+      xlab = "Number of Clusters",
+      ylab = "Within groups sum of squares")
```

结果如图14-26所示。

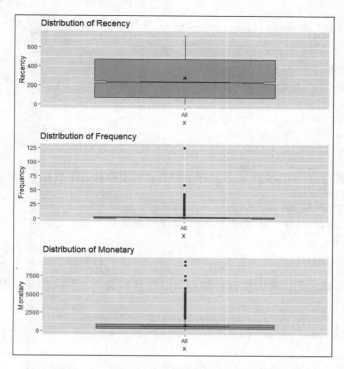

图 14-25　Recency、Frequency、Monetary 数据图

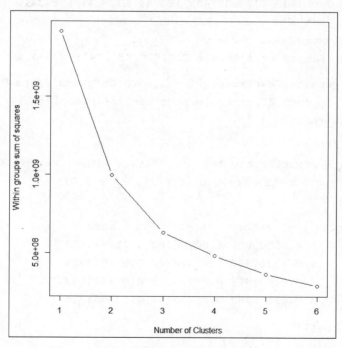

图 14-26　使用 k 均值聚类算法分客群为 1:6 群时的结果图

调用 NbClust() 函数评估，同样建议使用 k=4（读者可自行再评估）。

```
> NbClust(data=RFM_Model,distance="manhattan",min.nc=2,max.nc=6,
```

```
+           method='kmeans',index="scott")
$All.index
       2        3        4        5        6
10579.49 13310.34 17562.15 18899.31 20204.50

$Best.nc
Number_clusters    Value_Index
         4.000       4251.812
```

调用 kmeans()函数分为 4 客群并记录在 **RFMkc** 和 **RFMkcID** 数据框。

```
> kc <- kmeans(RFM_Model, 4)
> RFMkc=data.frame(RFM_Model,kc$cluster)
> RFMkcID=RFMkc
> RFMkcID$ID = RFM$ID
```

使用 k 均值聚类算法后,开始执行决策树,以便显示聚类规则(注意:设置的随机数不同,可能有不同的结果)。

```
> RFMkc$kc.cluster = factor(RFMkc$kc.cluster)
> str(RFMkc)
'data.frame':   4377 obs. of  4 variables:
 $ Recency   : num  432 471 629 402 123 44 130 341 69 43 ...
 $ Frequency : int  2 2 2 1 4 17 3 3 9 2 ...
 $ Monetary  : num  844 368 396 370 357 ...
 $ kc.cluster: Factor w/ 4 levels "1","2","3","4": 2 4 4 4 4 2 3 2 2 4 ...

> data_rpart<-rpart(RFMkc$kc.cluster~.,method='class',data=RFMkc,
+             cp=0.01,control=rpart.control(minsplit=2))
> summary(data_rpart)
Call:
rpart(formula = RFMkc$kc.cluster ~ ., data = RFMkc, method = "class",
    control = rpart.control(minsplit = 2), cp = 0.01)
  n= 4377

        CP nsplit rel error    xerror        xstd
1 0.7463807      0 1.00000000 1.00000000 0.017542122
2 0.2128686      1 0.25361930 0.25522788 0.011043948
3 0.0305630      2 0.04075067 0.04235925 0.004722583
4 0.0100000      3 0.01018767 0.01233244 0.002564726

> print(data_rpart)
n= 4377

> par(mar=rep(0.1,4))
> plot(data_rpart)
> text(data_rpart,cex=1.5)
> str(RFMkc)
```

```
'data.frame':    4377 obs. of  4 variables:
 $ Recency   : num  432 471 629 402 123 44 130 341 69 43 ...
 $ Frequency : int  2 2 2 1 4 17 3 3 9 2 ...
 $ Monetary  : num  844 368 396 370 357 ...
 $ kc.cluster: Factor w/ 4 levels "1","2","3","4": 2 4 4 4 4 2 3 2 2 4 ...
```

通过上述方法，利用决策树找出 4 个客户类别的规则，总计 6 条规则，如图 14-27 所示。

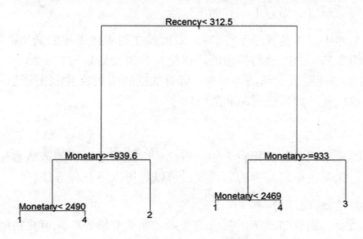

图 14-27　决策树分为 4 客群的图示

根据决策树的规则，可归纳出 4 个客户类别，分别为高价值顾客群、忠实顾客群、潜力顾客群和流失顾客群，其规则对照如表 14-1 所示。

表 14-1　顾客分群表

顾 客 群	规　　则
高价值顾客群	购买时间小于 312.5 日，平均交易金额大于等于 936.6 元，平均交易金额大于等于 2490 元
	购买时间大于等于 312.5 日，平均交易金额大于等于 933 元，平均交易金额大于等于 2469 元
忠实顾客群	购买时间小于 312.5 日，平均交易金额大于等于 936.6 元，平均交易金额小于 2490 元
	购买时间大于等于 312.5 日，平均交易金额大于等于 933 元，平均交易金额小于 2469 元
潜力顾客群	购买时间小于 312.5 日，平均交易金额大于等于 936.6 元
流失顾客群	购买时间大于等于 312.5 日，平均交易金额大于等于 933 元

我们可根据分析出的顾客类别及规则提供营销建议。

1．忠实顾客群

因为忠实顾客群就是较常来消费、买得较多的消费者，通过营销策略，达到让他们更"常"来买、买得更"多"的目的。通过提升品牌忠诚度使得消费者更频繁地来消费，可以使用以下两种方法：

（1）红利积点

每一次的消费金额可以按比例作为红利点数，可以提供下次消费的折扣。不但当次消费金额可能因为有之前的红利折扣而增加，也可以使回客率上升，提供消费者下次购买的诱因。

（2）以畅销商品作为主题

忠实顾客群的畅销商品通常为米、饼干、果蔬类、洗衣粉/液等家庭生活用品，可以不定期在超市内举办畅销商品的主题特卖。

2. 潜力顾客群

针对潜力顾客群对企业活跃度高但总金额不高的特性，提出的营销建议以促销高单价商品、刺激购货数量以提升总消费金额为主，有如下建议：

（1）消费者最易伸手拿与其视线平行的商品，因此在与视线平行的货架上放置高单价商品，即潜力顾客群的畅销商品：白兰地与葡萄酒，以提升其消费数量与总金额。

（2）将薄利多销的商品放在超商入口处，可有效提高购买数量，因此建议将此客群常购买的商品放置在超市入口处，如家庭用品和甜味饼干。

3. 流失顾客群

流失顾客群是不常来消费、买得较少的消费者，因此需通过营销策略增加顾客来超市的频率及提升消费金额。针对超市与品牌的忠诚度，可以利用以下三种方法：

（1）会员红利与满额赠活动

推出集点红利活动、满额赠的诱因吸引消费者（如消费满500元可兑换赠品或累计红利点数），来提高顾客单价与消费金额。

（2）以畅销商品作为主打

流失顾客群的畅销商品通常为米、鲜奶、家庭用品类商品，虽然所购买的金额较少，但仍与整体顾客所购买的商品类别相似。因此，仍需锁定流失客群喜好的商品进行促销提醒。

（3）观察竞争对手

除了商品因素外，流失客群的增加可能与附近邻近竞争者的兴起有关，可能为小型超市或其他连锁超市，超市店家应观察附近竞争者的类型，审视自我超市的类型、超市内的环境因素以及其他可改善的方向，用于加强与补足。

4. 高价值顾客群

高价值顾客的主要营销策略制定方向在于缩短与前一次购买时的时间间隔，提高购买的频率，使其成为具有高价值的忠实客户。

（1）回购礼

此策略主要用于减少高价值顾客购买的时间间隔，提升高价值顾客群的回购率。主要通过礼券或点数的发放来提高高价值顾客来超市消费的频率。

（2）系列活动

因为购买的物品大多与家庭必需品相关，因此通过系列活动（宠爱家人），提倡多项产品的组合优惠，并且拉长系列活动的时间，或者定期优惠不同商品组合，改善高价值顾客来超市的频率。

附录 A

安装 R

步骤 01   R 网站位于 http://www.r-project.org/。

首先去 R 语言的官网下载，单击左侧的 Download，选择 CRAN，或者单击右侧的 download R 链接，如图 A-1 所示。

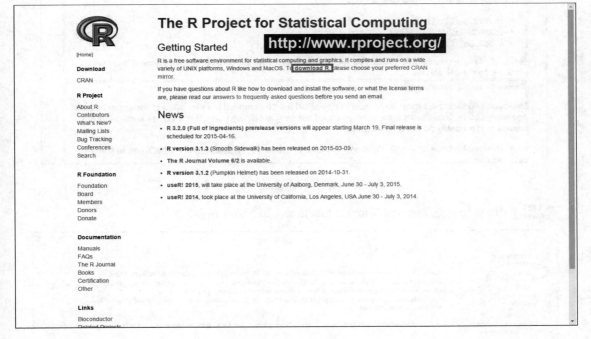

图 A-1   下载 R

步骤 02   然后选择相关的镜像，镜像是按照国家或地区进行分组的，找到 China，任意选择一个镜像即可，如图 A-2 所示。

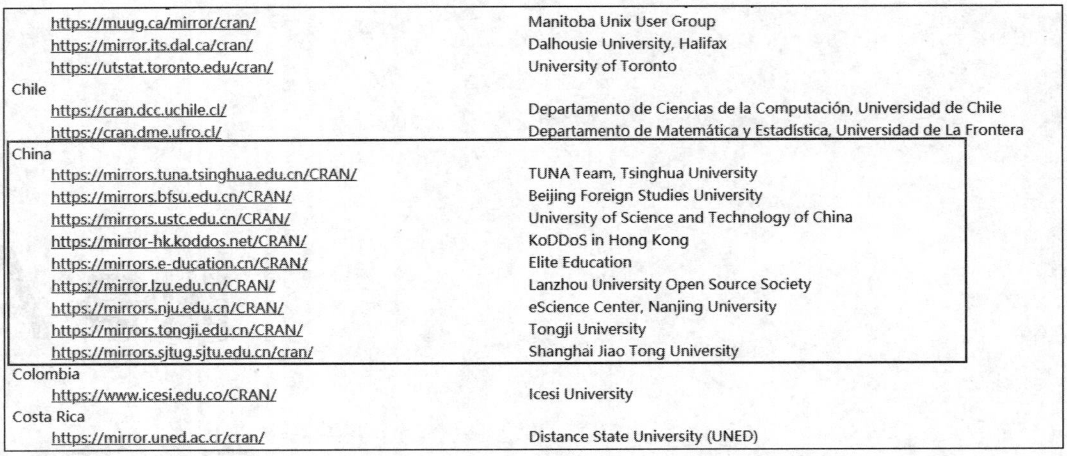
https://muug.ca/mirror/cran/
Manitoba Unix User Group
https://mirror.its.dal.ca/cran/
Dalhousie University, Halifax
https://utstat.toronto.edu/cran/
University of Toronto
Chile
https://cran.dcc.uchile.cl/
Departamento de Ciencias de la Computación, Universidad de Chile
https://cran.dme.ufro.cl/
Departamento de Matemática y Estadística, Universidad de La Frontera
China
https://mirrors.tuna.tsinghua.edu.cn/CRAN/
TUNA Team, Tsinghua University
https://mirrors.bfsu.edu.cn/CRAN/
Beijing Foreign Studies University
https://mirrors.ustc.edu.cn/CRAN/
University of Science and Technology of China
https://mirror-hk.koddos.net/CRAN/
KoDDoS in Hong Kong
https://mirrors.e-ducation.cn/CRAN/
Elite Education
https://mirror.lzu.edu.cn/CRAN/
Lanzhou University Open Source Society
https://mirrors.nju.edu.cn/CRAN/
eScience Center, Nanjing University
https://mirrors.tongji.edu.cn/CRAN/
Tongji University
https://mirrors.sjtug.sjtu.edu.cn/cran/
Shanghai Jiao Tong University
Colombia
https://www.icesi.edu.co/CRAN/
Icesi University
Costa Rica
https://mirror.uned.ac.cr/cran/
Distance State University (UNED)

图 A-2　选择镜像

步骤 03　选择操作系统。R 语言为 Linux、Windows 以及苹果系统都提供了相应的版本，如图 A-3 所示。

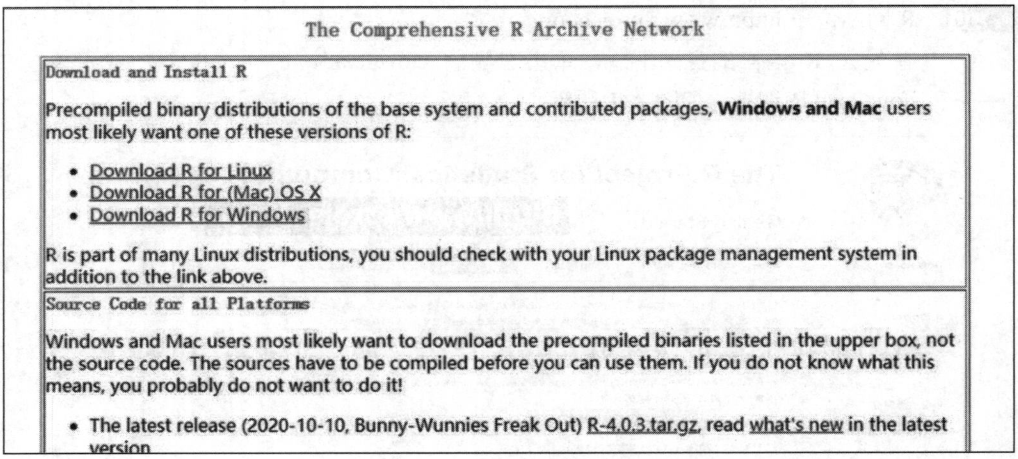
The Comprehensive R Archive Network
Download and Install R
Precompiled binary distributions of the base system and contributed packages, Windows and Mac users most likely want one of these versions of R:
Download R for Linux
Download R for (Mac) OS X
Download R for Windows
R is part of many Linux distributions, you should check with your Linux package management system in addition to the link above.
Source Code for all Platforms
Windows and Mac users most likely want to download the precompiled binaries listed in the upper box, not the source code. The sources have to be compiled before you can use them. If you do not know what this means, you probably do not want to do it!
The latest release (2020-10-10, Bunny-Wunnies Freak Out) R-4.0.3.tar.gz, read what's new in the latest version.

图 A-3　选择操作系统适用的 R 版本

步骤 04　单击 base 或者 install R for the first time，如图 A-4 所示。

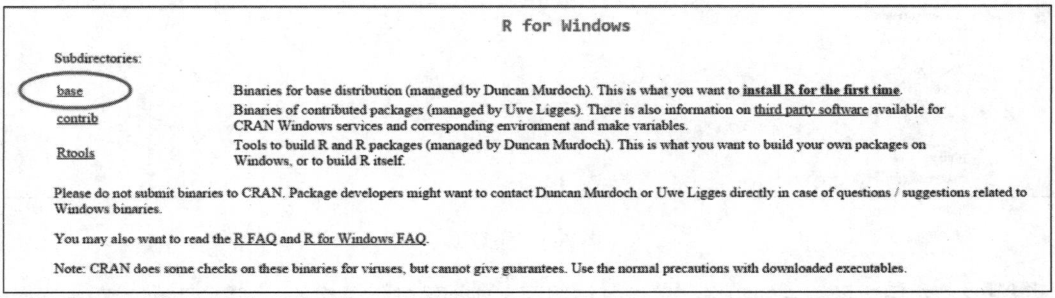
R for Windows
Subdirectories:
base
Binaries for base distribution (managed by Duncan Murdoch). This is what you want to install R for the first time.
contrib
Binaries of contributed packages (managed by Uwe Ligges). There is also information on third party software available for CRAN Windows services and corresponding environment and make variables.
Rtools
Tools to build R and R packages (managed by Duncan Murdoch). This is what you want to build your own packages on Windows, or to build R itself.
Please do not submit binaries to CRAN. Package developers might want to contact Duncan Murdoch or Uwe Ligges directly in case of questions / suggestions related to Windows binaries.
You may also want to read the R FAQ and R for Windows FAQ.
Note: CRAN does some checks on these binaries for viruses, but cannot give guarantees. Use the normal precautions with downloaded executables.

图 A-4　选择 base 或单击链接

步骤 05 单击 Download R *** for Windows 的链接就能下载 R 了，如图 A-5 所示。我们将文件下载到本地计算机上。

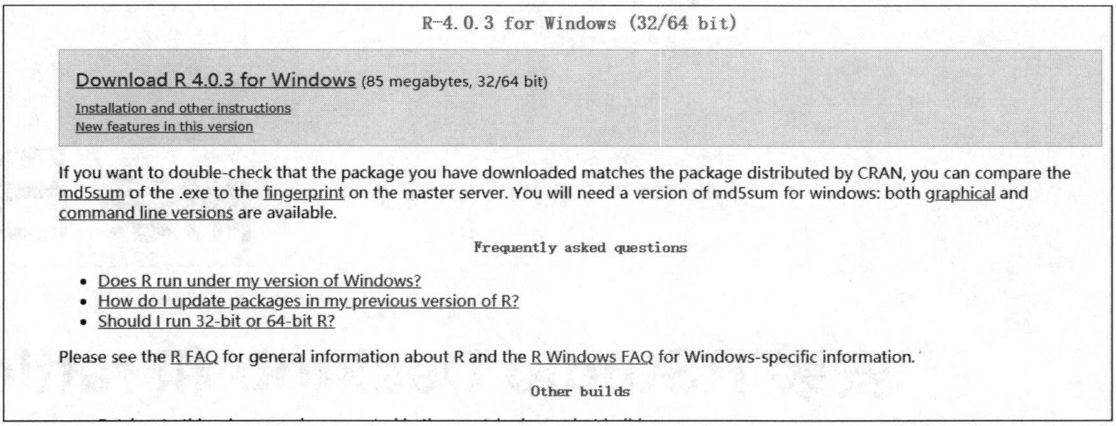

图 A-5　单击下载链接

步骤 06 双击 EXE 文件进行安装。首先选择语言，这里选择"中文（简体）"即可，然后单击"确定"按钮，如图 A-6 所示。默认选择下一步，选择安装路径，再单击下一步。安装完成后，我们发现在桌面上出现了 R 的图标，双击即可启动，而后出现如图 A-7 所示的界面，证明 R 安装成功了。

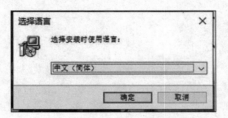

图 A-6　选择"中文（简体）"

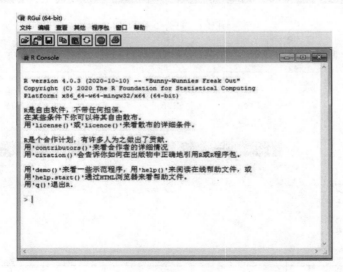

图 A-7　R 软件界面

附录 B

安装 RStudio Desktop 和 rattle

步骤 01 首先前往此网站 http://www.rstudio.com/ 下载 RStudio Desktop，如图 B-1 所示。

图 B-1 打开下载网站

步骤 02 进入页面后，单击 Free 下面的 DOWNLOAD 按钮，如图 B-2 所示。

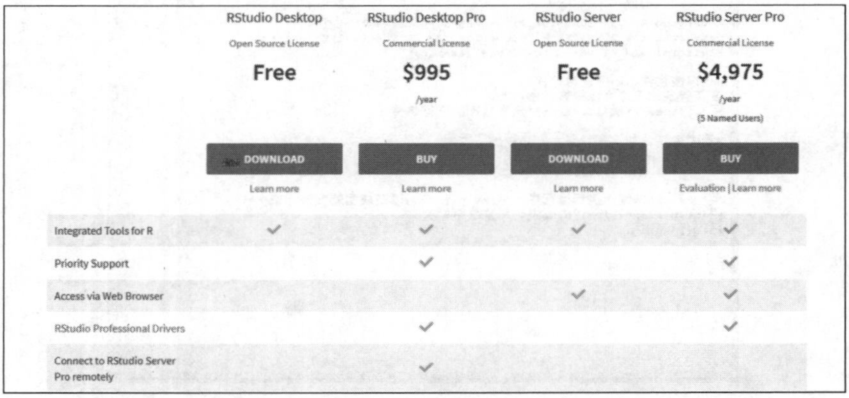

图 B-2 单击 DOWNLOAD 按钮

步骤 03　进入如图 B-3 所示的页面，下载安装文件。注：1.安装 RStudio 之前需要安装 R 3.0.1+；2.单击下载 Requires Windows 10/8/7 (64-bit)。

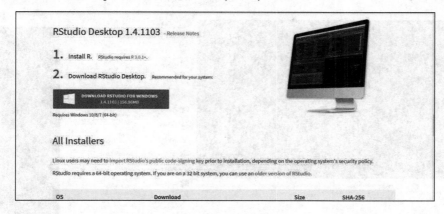

图 B-3　下载安装文件

步骤 04　安装步骤。双击安装文件进入欢迎界面，单击"下一步"按钮，如图 B-4 所示。然后选择安装目录，单击"下一步"按钮，如图 B-5 所示。

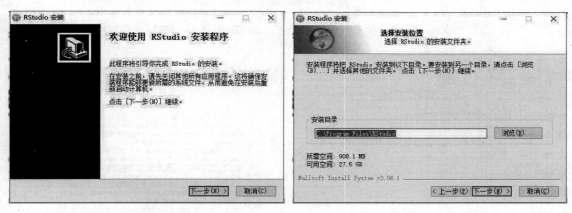

图 B-4　欢迎界面　　　　　　　　　　图 B-5　选择安装目录

单击"安装"按钮，如图 B-6 所示，安装完成并生成桌面快捷方式。

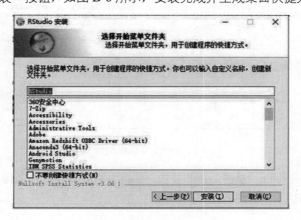

图 B-6　单击"安装"按钮

单击 Tools 下面的 Global Options 选项，在 Options 界面选择 Appearance 选项，编辑界面字体，如图 B-7 所示。

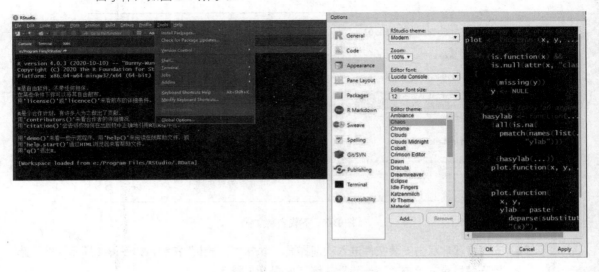

图 B-7　编辑界面字体

单击 Packages 选项，改变镜像地址为 China（Beijing 1）（这样下载速度会快一些），如图 B-8 所示。

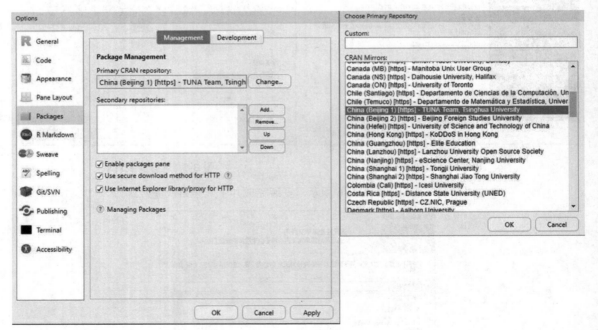

图 B-8　改变镜像地址

接下来安装 rattle。单击 Install 按钮，出现 Install 界面后分别输入 RGtk2 和 rattle，再单击 Install 按钮，如图 B-9 所示。

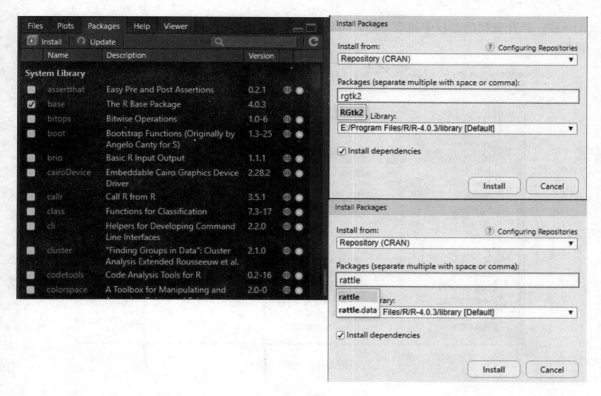

图 B-9　安装 rattle

然后在 Packages 选项卡中勾选 rattle 和 RGtk2 复选框，在代码区输入 rattle()，按回车键即可，如图 B-10 所示。

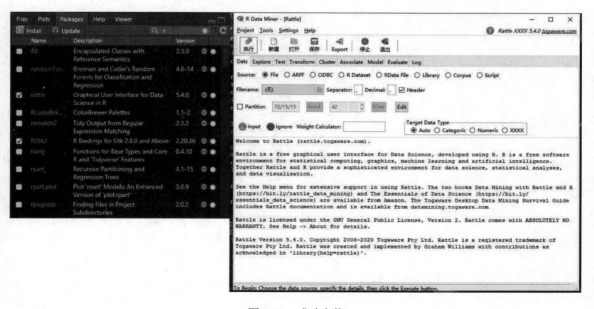

图 B-10　成功安装

注：若在安装的过程中出现问题，可以将这个 gtk 文件放入如图 B-11 所示的路径中进行解压。

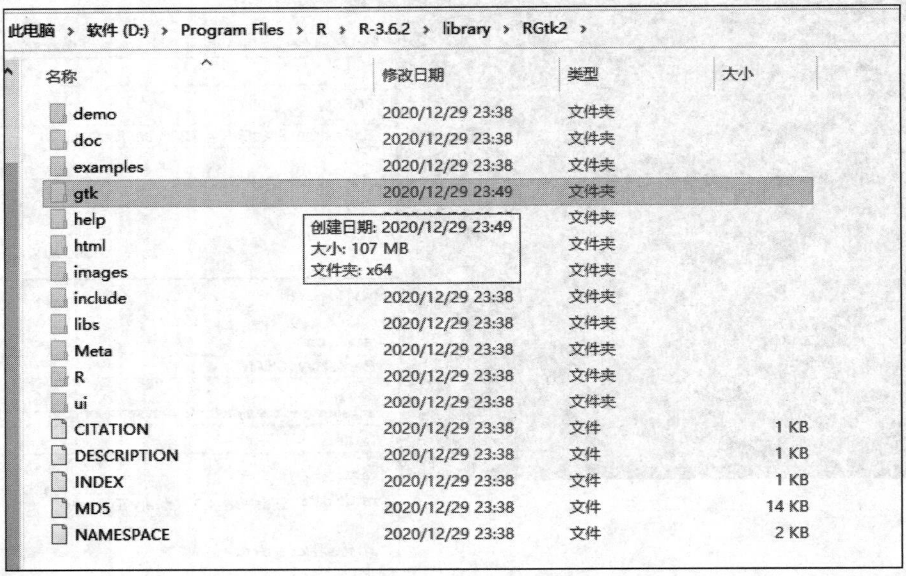

图 B-11 解压路径

附录 C

R 语言指令及用法

指　　令	功　　能
基本操作	
demo(包)	演示包的功能
library(包)	引用或加载包
data(数据)	引用或加载数据
head(数据,n)	显示数据前 n 笔数据，默认 n=6
tail(数据,n)	显示数据后 n 笔数据，默认 n=6
setwd("路径")	切换工作目录
getwd()	取得工作目录
length()	计算对象中元素的数量
mode()	获取对象的数据类型
class()	获取对象的类
str()	获取对象的数据结构
rm(list=ls());gc()	清除所有对象
对象及其运算	
%/%	整除
%%	求余数
%*%	矩阵相乘
%in%	判断是否在某个集合内
|	或（OR）
&	且（AND）
!	否（NOT）
seq(from, to, by)	产生以 by 为递增值的向量
assign("对象名称", 表达式)	创建对象，并将表达式的值传入
c()	创建向量对象

（续表）

指　　令	功　　能
对象及其运算	
array(x, dim=c())	按照 dim 维创建 array 对象
matrix(x, nrow, ncol, byrow)	创建 x 为 nrowr × ncol 的矩阵，byrow=T 时以行的顺序排列
factor(x, levels=c())	按照 levels 排列创建 factor 对象
data.frame()	创建数据框（data frame）对象
list(name1=value1,…)	创建列表（list）对象
rbind()	按行合并
cbind()	按列合并
t()	转置矩阵
det()	求矩阵行列式的值
eigen()	计算矩阵特征值和特征向量
solve()	求反矩阵
solve(A, b)	求矩阵 Ax=b 的 x 解
qr()	求矩阵 qr 分解
svd()	求矩阵 svd 分解
edit()	以电子表格方式编辑对象数据
view()	以电子表格方式显示对象数据
fix()	以电子表格方式修改对象数据
read.table()	输入多种格式的数据文件
scan()	输入数据
read.csv()	输入逗号分隔的 CSV 格式文件
write.table()	输出多种格式的数据文件
write.csv()	输出逗号分隔的 CSV 格式文件
save()	将对象存储为 Rdata 格式
load()	加载 Rdata 文件内的所有对象
file.choose()	以窗口选取来替代路径，搭配输入函数
as.	各种对象的转换
流程控制	
ifelse(condition, T, F)	用于二分类逻辑判断，condition 成立时执行 T，否则执行 F
if(condition){表达式 1} else{表达式 2}	condition 成立时执行表达式 1，否则执行表达式 2
switch(计算值,表达式 1，表达式 2,…)	按照计算值（整数或文字）决定要执行的表达式
while(condition){表达式}	condition 成立时执行表达式
repeat{}	重复执行表达式直到跳出（break）
break	终止并跳出循环
next	跳过当前循环体其后尚未执行的语句，直接执行下一轮循环
apply(x, MARGIN, function)	对矩阵或数据框架的 x 对象按列或行（MARGIN=1 或 2）执行 function 函数

（续表）

指　　令	功　　能
流程控制	
lapply(x ,function)	对 x 对象执行 function 函数并以列表方式返回
sapply(x, function)	对 x 对象执行 function 函数并返回较简单的向量或矩阵方式
数学函数	
sum()	返回元素总和
abs(x)	返回 x 绝对值
sqrt(x)	返回 x 平方根的值
ceiling(x)	返回大于等于 x 的最小整数
floor(x)	返回小于等于 x 的最大整数
round(x,n)	将 x 四舍五入到第 n 位
trunc(x)	返回 x 的整数部分
max()	返回最大值
min()	返回最小值
sign()	判断正负号
exp()	指数函数
log(x,base)	对数函数
sin()、cos()、tan()、asin()、acos、atan()	三角函数、反三角函数
mean()	返回平均值
median()	返回中位数
var()	返回方差
sd()	返回标准差
range()	返回最大值及最小值
IQR()	返回 IQR 值
cor(x, y)	返回 x 及 y 的相关系数
绘图函数	
plot(x)	以序号为横坐标（x 轴）、y 为纵坐标（y 轴）来绘图
plot(x, y)	以 x（x 轴）和 y（y 轴）来绘图
pie(x)	绘制饼图
boxplot(x)	绘制盒形图
stem(x)	绘制茎叶图
dotchart(x)	绘制散点图
hist(x)	绘制直方图
barplot(x)	绘制条形图
contour(x, y, z)	绘制等高线图
points(x, y)	加上点
lines(x, y)	加上直线
text(x, y, labels)	在指定位置显示指定文字
abline(a, b)	加上直线（y=ax+b）

（续表）

指　　令	功　　能
绘图函数	
abline(h=y);abline(v=x)	加上水平线或垂直线
legend(x ,y, legend)	在指定位置画出图例
title(main, sub)	加上主标题或副标题
locator(n, type)	选择当前图形上的特定位置，最多取 n 次
identify(x, y, n, label)	在指定点旁显示其在原向量中的指标值，最多选择 n 次
par(margin=c(a, b ,c, d))	设置离底部、左边、上方及右边的边界值，单位为分米
par(mfcol=c(a, b))	以 a×b 矩阵将多张图形画在同一页，按行的顺序画图
par(mfrow=c(a, b))	以 a×b 矩阵将多张图形画在同一页，按列的顺序画图
其他函数	
na.omit();na.exclude()	删除 NA 值
is.na()	判断元素是否为 NA
na.rm=T	在某些函数内使用，可删除 NA 值
proc.time();system.time()	测量程序代码运行的时间
system("command")	调用系统函数
Sys.time()	读取时间
Sys.sleep(x)	让程序暂时停止执行 x 秒